Whales & Dolphins
of the World

Whales & Dolphins of the World

MARK SIMMONDS

photography by seapics.com

This edition published in 2013 by New Holland Publishers
London • Cape Town • Sydney • Auckland
www.newhollandpublishers.com

Garfield House, 86–88 Edgware Road,
London W2 2EA, UK
Wembley Square, First Floor, Solan Road,
Cape Town, 8001, South Africa
Unit 1, 66 Gibbes Street, Chatswood, NSW 2067, Australia
218 Lake Road, Northcote, Auckland, New Zealand

A CIP catalogue record for this book is
available from the British Library.

ISBN 978 1 78009 461 8

Publisher Simon Papps
Editor Sue Viccars
Design: Lorena Susak
Production Olga Dementiev

Photograph on page 53 © Simon Papps

Printed and bound in China by Toppan Leefung Printing Ltd

Image details
FRONT COVER: Beluga Whale (above) and Pacific
White-sided Dolphins (below).
BACK COVER: Humpback Whale.
PAGE 1: Short-beaked Common Dolphins.
PAGE 2: Humpback Whales blow.
OPPOSITE: Humpback Whales breaching.
PAGE 6, from top to bottom: Humpback Whales
lunge-feeding; Grey Whale juvenile in kelp forest;
Bottlenose Dolphin.
PAGE 7, from top to bottom: Grey Whale subadult;
Atlantic Spotted Dolphins; Beluga Whale.
PAGE 153: Mother and calf Humpback Whales.
PAGE 154: Bottlenose Dolphin.
PAGE 160: Orcas spy-hopping.

Contents

Introduction

THE FASCINATION OF WHALES AND DOLPHINS

This book is intended as both a celebration of the whales and dolphins of the world and an introduction to their diversity, biology and conservation. Human beings have a long-standing fascination with these animals, and we have interacted with them in many different ways over the centuries. Today they generate huge interest, as evidenced in September 2003 when the BBC programme *50 Things To Do Before You Die* presented the results of a survey of 20,000 members of the UK public. At the very top of the list – as the most popular activity of all – was 'swimming with dolphins'; not far behind, at number four, was 'whale-watching'.

Stepping back in time, the Greek philosopher Aristotle (c. 384–322 bc) recognized that whales and dolphins are air-breathing mammals that bear live young. Unhappily, the significance of this knowledge was largely forgotten over the following centuries. Indeed, it was not so long ago that the popular image of whales was of giant marine monsters to be feared by sailors. Early illustrations of whales – including those in the first modern natural history accounts, around the mid-16th century – reflected

this theme, depicting, for example, fantastic struggles between vast whales and monster giant squid. Some classical legendary monsters, such as the sea serpent, the mermaid and the unicorn, may have been based on experienc es with the real animals that we now know as whales, dolphins and porpoises.

The monster image was also perpetuated in Herman Melville's popular epic novel Moby Dick. Still, arguably, the best-known whale story, Melville's book was actually inspired by the real events that started on 20th November, 1820, when the whaling vessel *Essex* from Nantucket in the USA was holed by a whale. Working from open rowing boats, the crew had harpooned three Cachalots (Sperm Whales). The largest, a 26-metre (85-foot) specimen, responded by ramming the main vessel, causing it to sink two days later and casting the crew adrift on the high seas.

The historical view of Cachalots and other large whales as dangerous monsters contrasts greatly with the current enthusiasm for observing them in close proximity in their natural habitat, an activity that exhibits the considerable trust that humans place in their peaceful natures. People are prepared to board relatively small boats to marvel at (rather than fear) the large size of these animals and to witness their behaviour at close quarters.

The chapters that follow divorce the monsters of the myths from the equally fascinating realities of these animals. I will invite you to appreciate the surprising diversity of whales, dolphins and porpoises, their remarkable biology and life strategies. I will also ask you to consider the threats that they face, and how these could be addressed. This book is only possible because of the efforts of many dedicated scientists who have

OPPOSITE   A group of amiable Atlantic Spotted Dolphins off the Bahamas.
BELOW   Another kind of marine mammal, a Polar Bear, feeding on the carcass of a Bowhead Whale (Alaska).

worked hard, often over many decades and under the most arduous conditions, to improve our understanding of these often-elusive animals. Thanks to their work, recent years have witnessed the publication of some very important research that has hugely improved and changed our perceptions. Nonetheless, many mysteries remain and our knowledge of the behaviour and societies of the cetaceans still lags many decades behind our appreciation of other key mammals, such as our close relatives the primates.

Royalties from *Whales and Dolphins of the World* will go to the Whale and Dolphin Conservation Society (WDCS), a charity dedicated to the conservation and welfare of the cetaceans. As will be seen in the concluding chapters, protecting these 85-plus species and their environment, as humankind increasingly encroaches into their habitats, presents major challenges and WDCS

runs projects all over the world to help meet these.

I hope that you will enjoy this book and that it will help to convince you that we should make every effort to share our planet in good fellowship with these wonderful animals.

Mark P. Simmonds

Director of Science,
Whale and Dolphin Conservation Society

ABOVE Pacific White-sided Dolphins, shown here, are powerful swimmers and are usually found in deep offshore waters.
OPPOSITE A Humpback Whale in the waters off Hawaii. Knobbly lips, white underbelly and long flippers are characteristic features of this ocean giant.

Chapter One

WHO ARE THE WHALES AND DOLPHINS?

Oceans cover 70 per cent of the surface of our planet. In this vast expanse of water and in some of the larger rivers of Asia and South America, there exists a fascinating and mysterious group of animals – cetaceans – about which humankind still knows very little. Many species lead lives that are mainly hidden in the depths or far out to sea, rarely, if ever, coming into contact with humans. But a few, including the Bottlenose Dolphins, sometimes associate with us voluntarily. While whales, dolphins and porpoises are generally fish-like in appearance, they are actually mammals and exhibit a diversity of adaptations that have enabled different species to colonize a wide variety of marine habitats and ecosystems.

In this first chapter we will start to explore the diversity of whale-kind, including the remarkable range of species that has evolved and with which we still share this planet.

Introduction to the cetaceans

We have just had the most incredible and inspiring sighting of a mighty whale… and a description that doesn't in some way invoke the mystical quality of this event will not convey the experience. It is Monday and we have just entered the deep waters beyond the continental shelf edge in the third leg of our survey block in the waters to the west of the Orkney and Shetland Islands. The swell is less than one metre, the wavelets small, the day overcast and, after several hours of seeing nothing, we sight high blows ahead. Then, surging through the waves towards the ship, casting the water aside with a power and nonchalance unknown anywhere else in the animal kingdom, comes the whale. She is travelling at the surface, powerfully pushing the water aside, her blows leaving vapour and spray in the air. She passes within a few hundred metres of the prow, crossing to our starboard side. We see her back clearly above the waves, a triangular fin and her blowholes. Off the starboard side, she turns away from us and, for one glorious timeless moment, we look along the back of a fifty-foot whale, past her fin and into the briefly open caverns of her twin blowholes, as she takes a great breath.

Extract from the author's journal during a cetacean survey in the waters to the north and west of Scotland, 30th July, 1998.

There are at least 85 species of whales, dolphins and porpoises, and they all belong to the order of mammals known as Cetacea: hence, cetaceans. There is a far greater diversity of form and function amongst cetaceans than is commonly appreciated. Cetaceans have evolved to exploit habitats ranging from large, freshwater river systems, through the surface waters of coastal and mid-ocean regions, into the abyssal depths of the oceans. Every year reveals fascinating discoveries about their life histories, social behaviour and habitats as new research techniques are developed. Yet many species – and much cetacean natural history – are only poorly understood, even today.

The cetacean order includes the real giants of the animal kingdom, such as the Blue, Fin and Humpback Whales, the highly social dolphin species, the diminutive porpoises, the Arctic-dwelling white whales, the elusive beaked whales and many others. They are all warm-blooded, give birth to live young and breathe air, sharing many attributes with terrestrial mammals, including humans. New species are still being discovered – especially in the little-known beaked whale group – and new genetic techniques are revealing that some recorded species are actually two or more.

Cetaceans have evolved to live in an aquatic environment that is significantly more complex and three-dimensional than the world we understand. Their habitats are fixed to varying degrees. Many species of beaked whales are found over deep ocean trenches where they prey on fish or squid that live in the depths of the sea. Other cetaceans, including many dolphin species, exploit more transient resources associated with particular water characteristics such as temperature, chemistry or clarity.

Different conditions favour particular prey animals and possibly cetaceans themselves. There is a strong correlation between cetacean distribution and water temperature on both a global and a more local scale. Some species show seasonal changes in habitat use, including those that migrate huge distances between their warm winter breeding grounds and cold summer feeding areas close to the poles.

OPPOSITE A Bryde's Whale feeding side by side with Common Dolphins. The huge Bryde's Whale, seen here lunging from the water with a mouth full of prey, is a filter-feeding baleen whale.
RIGHT A group of distinctively marked Southern Right-whale Dolphins off the coast of Chile.
PAGE 12 An Orca or Killer Whale – actually the largest dolphin species – breaches off San Juan Island.

ABOVE Pantropical Spotted Dolphins riding in the wake of a vessel off the coast of Hawaii.

EVOLUTIONARY DEVELOPMENT

The evolutionary history of cetaceans started after the dinosaurs declined, some 50 million years ago, with a mammal that resembled an otter (except for its little hooves) and which hunted fish in the Tethys Sea, in what is now the Mediterranean region. The fossil record reveals that the descendants of this hairy mammal became increasingly aquatic and, over time, ceased using their limbs to swim in favour of their tails. Eventually their forelimbs became flippers, and all external traces of the hindlimbs disappeared. However, inside the front flippers of modern whales and dolphins is an arrangement of bones very like those in a human hand (see page 71). Most of the cetacean genera that we know today were already present around 10 million years ago.

These evolutionary changes – moving away from the

ABOVE   A spectacular Spinner Dolphin exhibiting the cause for its name by leaping from the water and simultaneously twirling around its long-axis in Hawaii.
RIGHT   The largest animal ever to have lived accompanied by some smaller relatives: a Blue Whale is flanked by Risso's Dolphins off California. The whale has just exhaled a mighty jet of water vapour from its blowholes.

basic anatomy of a terrestrial mammal – make perfect sense for animals adapting to an aquatic lifestyle. However, this does not explain the impressively large size of some cetaceans. The Blue Whale is not just the largest animal alive today, it is also – along with its close relative, the Fin Whale – probably the largest animal ever to have existed. The Fin Whale reaches a maximum size of about 27 metres (89 feet) and weighs some 125 tons; the Blue Whale reaches about 34 metres (110 feet) and weighs around 190 tons. These are truly massive animals and they have evolved to such great body size aided by the fact that the aquatic medium in which they live helps to support their weight.

Large size may be a good adaptation for survival in cold water. The larger the animal, the lower its relative body surface area. The lower the relative body surface area exposed to cold water, the less the relative area for heat loss. However, many smaller cetaceans are also found in cold polar waters and they possess other adaptations that enable their survival, including the thick blanket of fat, or blubber, just under the skin, which acts as an important insulator.

During the spectacular evolution of cetaceans, organs that typically protrude in land mammals (ears, mammary glands and male genitalia) all became internalized. A smooth skin developed, and eyes and ears became adapted for use underwater. During this process the skull changed shape quite considerably, allowing the nostrils of cetaceans to migrate to the top of their heads to form one or two 'blowholes'.

Taxonomy

The classification of cetaceans is still rapidly evolving. Biologists may seem a little obsessed with classifying animals and naming species, but this cataloguing is an important step in understanding and conserving the natural world. The modern cetaceans – order Cetacea – can be divided into two suborders: the Odontoceti or 'toothed cetaceans' (which includes all dolphins and porpoises) and the Mysticeti or 'baleen whales' (which typically feed by filtering planktonic crustaceans or small fish through sieve-like structures – baleen plates – in their huge mouths). Baleen whales are also characterized by having two blowholes on top of their heads, instead of the single blowhole possessed by the toothed cetaceans (or odontocetes).

Cetaceans are further divided into families of closely related species. All the beaked whale species, for example, are in the family Ziphidae, indicating that they are all related to each other – with a common ancestor.

Within families, two-part scientific names, based on the Latin language, are given to individual species. For example, *Tursiops truncatus* is the name tag for the Common Bottlenose Dolphin. The recently defined *Tursiops aduncus* is the Indo-Pacific Bottlenose Dolphin. The generic name *Tursiops* reflects the fact that they are very closely related. The second part of the name is the distinct species name. There can also be a third part to the scientific name where subspecies have been identified. For example, *Tursiops truncatus ponticus* has been proposed for the distinctive Black Sea Bottlenose Dolphin.

The present count of cetacean species is 85 and still climbing (a full list is provided on pages 150–151).

The suborder Odontoceti contains ten families:
- **Phocoenidae** – porpoises.
- **Pontoporiidae, Platanistidae, Lipotidae, Iniidae** – the four families of river dolphins.
- **Monodontidae** – better known as the 'white whales'.
- **Ziphidae** – the elusive beaked whales.
- **Physeteridae** – currently containing only one species, the giant Cachalot or Sperm Whale.
- **Kogiidae** – the Dwarf and Pygmy Sperm Whales.
- **Delphinidae** – the family of the true dolphins.

The suborder Mysticeti, the baleen whales, contains far fewer species (at least 14), six genera and only four families:
- **Eschrichtiidae** – the Grey Whale.
- **Balaenidae** – the Bowhead Whale and right whales.
- **Neobalaenidae** – the Pygmy Right Whale.
- **Balaenopteridae** – the rorqual whales.

ABOVE The top of the head of a Bryde's Whale showing the pair of blowholes characteristic of a mysticete or baleen whale.
OPPOSITE A Common Bottlenose Dolphin – the best known of the dolphin species with its distinctive beak, prominant dorsal fin and typical dolphin body shape. This sociable individual was photographed off the Cayman Islands in the Caribbean with colourful tube sponges in the foreground.

The porpoises

There are six living species of porpoise: Burmeister's, Harbour, Dall's, Spectacled and Finless Porpoises and the Vaquita. All are relatively small, less than 2.5 metres (8 feet) in length, robust and lacking a beak. Dall's Porpoise and the little-studied Spectacled Porpoise are primarily offshore animals. The Harbour Porpoise has been recorded quite a distance from the shore but is regarded as a primarily coastal species, as are the Burmeister's and Finless Porpoises and the Vaquita. The Finless Porpoise can also be found quite far into some river systems. Compared to other cetaceans, porpoises mature at an early age, have a relatively high reproductive rate (some breed each year and are capable of concurrent pregnancy and lactation), but rarely live longer than 20 years. These relatively short lives and high reproductive rates reveal life strategies significantly different to those of many other cetaceans.

We probably know more about the Harbour Porpoise than the other species: A newly born Harbour Porpoise weighs just 5 kilograms (11 pounds) and is about 70 centimetres (28 inches) long. The mothers nurse their calves for up to a year: a relatively short time for a cetacean. Breeding seems to be strongly seasonal and porpoise calves usually arrive in the summer, when water temperatures are most benign. The calves grow rapidly and are soon wrapped in a thick layer of insulating blubber. Porpoises spend much of their time feeding, and need to find about eight per cent of their body weight each day.

Unfortunately, one species of porpoise may soon become extinct. The Vaquita's distribution is limited to the shallow upper Gulf of California, in Mexico, and despite heroic efforts to conserve it, it is one of the most endangered mammals on earth. Vaquita, rather unflatteringly, means 'little cow'. It is probably the smallest of all the cetaceans (although there are several cetacean species of similar size) reaching a length of only about 1.5 metres (5 feet). Its biology is not well known, but seems similar to that of the Harbour Porpoise.

There are no more than a few hundred Vaquitas left in their shallow water habitat. The primary problem is intensive fishing activity. Vaquitas are accidentally caught and die in trawl and gill nets, some of which are set for a large fish, called the Totoaba, which is also endemic to the upper Gulf. In the mid-20th century, the Totoaba was plentiful and supported a large fishery but now, following over-exploitation, it too has become endangered. The porpoise and the fish will both become extinct unless the fisheries are speedily controlled.

Dall's Porpoise is another typical stocky porpoise, dark grey with a distinctive white flash on its rear flank. It is unusual in being fast-swimming and highly active at the water's surface. Most porpoises breach rather gently, scarcely making any wake, but the Dall's Porpoise speeds through the water making characteristic 'rooster tails' of spray. Despite its speed, it is one of the porpoise species that is subject to major hunting. Its North Pacific range brings it close to the shores of Japan. A 'hand-held harpoon fishery' operates off Hokkaido in the late spring, summer and autumn, moving in the winter to the north coast of Honshu. The porpoises are taken for human consumption, and are brought ashore, where their meat is usually then distributed via markets. The meat from one Dall's Porpoise is worth some 20,000 Japanese yen (about £106 British pounds, or $169 US dollars). However, whilst Japanese officials defend these hunts in the name of tradition, it appears that most Japanese people have no idea that they occur, and whale and dolphin meat is no longer an important staple food.

In the mid-1980s the annual catch of Dall's Porpoises stood at around 10,000, but this rose to over 40,000 in 1988. Despite an initial reduction in take – in response to concern expressed by the International Whaling Commission (IWC) – the hunt has subsequently increased. The 1997 take was 18,000 porpoises, and there was also a bias in the number of lactating female porpoises killed. Hunters exploit the fact that females will not leave their slower-moving calves, which can become exhausted by the chase, making both mother and calf easier to catch. Taking females with calves will speed up the decline of the populations. In recent years the number of Dall's Porpoises taken annually has declined, perhaps due to reduced demand (their meat has been found to contain high levels of mercury) or perhaps due to reduced numbers of porpoises. The Dall's Porpoise hunt is one of a number of hunts of 'small cetaceans' in Japan that are causing concern, particularly because they are poorly regulated. Burmeister's Porpoise is hunted in Peru and Chile.

All porpoises living in coastal areas are affected to some degree by human activities. The Harbour Porpoise seems to be especially vulnerable to gill nets set on the seabed, where it feeds, and many thousands are killed each year in the North and Celtic Seas. It is also already critically endangered in the neighbouring Baltic Sea and is rare to the south of the UK.

The endangered Yangtze River Porpoise population (actually a subspecies) of the Finless Porpoise is also in trouble, and is exposed to the same problems that are driving the river dolphins towards extinction (see page 22).

ABOVE A Harbour Porpoise in the Sognefjord in Norway. The Harbour Porpoise is still a relatively familiar sight in some North Atlantic coastal waters, although less so than it used to be. It is usually seen in small groups of two to three or as solitary individuals
OPPOSITE A Dall's Porpoise in the Bering Sea.
BELOW The body of a Burmeister's Porpoise killed in a fishery in Peru. The small fins and absence of a prominent beak are typical of all porpoises.

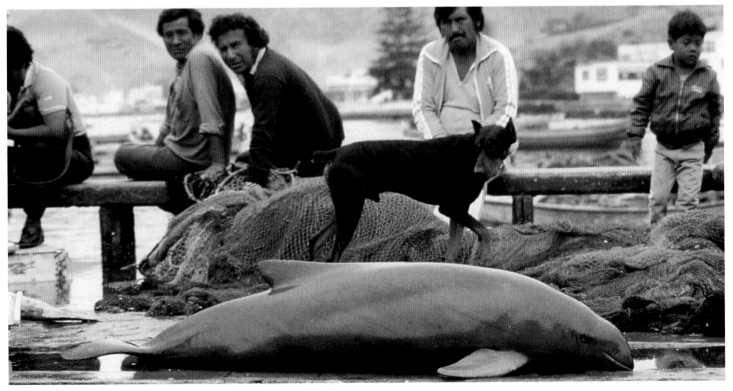

The dolphins

The River Dolphins

If you tend to swim on your side, with one large flipper reaching down to touch the sediment, and you have a markedly long beak and very small eyes, then you are probably a river dolphin. The two subspecies of the South Asia River Dolphin – the Ganges (or Gangetic) River Dolphin and the Indus River Dolphin – are included in the family Platanistidae. In the family Iniidae are the Boto (also called the Pink River Dolphin) and its three subspecies: which are also known as the Orinoco, Amazon and Bolivian River Dolphins. The Baiji (or Yangtze/Chinese River Dolphin) has the family Lipotidae to itself, and the Franciscana (or La Plata River Dolphin) is in yet another family: the Pontoporiidae. The number of distinct families reflects the fact that many of the river dolphins are only distantly related: several ancestral lines converged to produce these modern species, which resemble each other in appearance and mostly live in tropical rivers.

The Franciscana is rather an aberrant river dolphin because it lives in the sea, but it is very similar in many ways to the other species. It is only found in the shallow coastal waters of the Atlantic coasts of Brazil, Uruguay and Argentina. The ranges of the Ganges and Indus River Dolphins include a number of larger river systems in India, Pakistan, Bangladesh, Nepal and Bhutan. The Baiji, meanwhile, formerly occurred along some 1,600 kilometres (1,000 miles) of the Yangtze River in China (see below). Finally, the Boto is found in the extensive river systems that ramify through Venezuela, Colombia, Ecuador and Brazil.

The river dolphins have a distinctive appearance and share features that are adaptations to life in a fast-moving environment. They all have characteristically small eyes, although – with the exceptions of the Indus and Ganges Dolphins, which may see little more than light and dark – they have good vision. River dolphins all have broad flippers and very long slender beaks with many small pointed teeth that are rather similar to the facial arrangement of some crocodiles.

The Baiji was declared 'functionally extinct' in 2006, although a few individuals may survive. Pollution and the damming of the various tributaries where it lived have

brought about its demise. It has also been hunted for meat and other products in the past, but in the late 20th century other problems included capture in fishing nets, collisions with fast-moving vessels and damage from underwater blasting during construction work. Its habitat is shared with a subspecies of the Finless Porpoise, which is also declining rapidly. These animals occur within the territory of a single country, China, which therefore has the responsibility to take action to conserve them. However, it seems that the survival of these animals is incompatible with the demands of a modern and expanding society.

The river dolphins of the Indian subcontinent are also classified as endangered. The Indus River Dolphin population now numbers around 1,000 individuals, and the Ganges River Dolphin numbers a few thousand. The problems these species face are very similar to those of the Baiji: habitat fragmentation, pollution, disturbance and

occasional hunting. The Indus River Dolphin is now only found in the main channel of the river, although previously it inhabited many side branches. Irrigation dams have divided its population, and isolated sub-populations have vanished. The situation in the Ganges is similar; more than 50 dams have been constructed along this river system and more are planned, further dividing up the available habitat.

Whilst the Boto is still relatively widely distributed through much of the Amazon and Orinoco River systems, the species faces a range of serious threats nonetheless. Botos are sometimes hunted deliberately, and there is

ABOVE The endangered Ganges or Gangetic River Dolphin (also known as the Susu) surfacing near a fisherman in one of the Indian river systems where it can still be found.

OPPOSITE The possibly extinct Baiji or Yangtze/Chinese River Dolphin holding a fish in its slender jaws.

ABOVE Botos, which are also known as Pink or Amazon River Dolphins, are well adapted to life in a fast-moving environment and possess a sophisticated echolocation system.

BELOW There is evidence that some river dolphins, such as this Brazilian Boto, co-ordinate their activities to herd fish but the biology of most river dolphins is still little understood.

expanding use of their carcasses for fish bait. They may also be killed by fishermen, in retaliation for perceived competition for increasingly scarce fish stocks. Moreover, the growth of human populations along these river systems, combined with various dam projects, mean that, in due course, the Boto could face the same fate as the other river dolphins. There is also growing concern about pollution of their habitat by mercury and other heavy metals, resulting from gold and other mining operations, as well as pesticide and herbicide run-off from cash crops.

The 'marine' river dolphin, the Franciscana, has similar conservation problems. The narrow strip of coastal waters that it inhabits is nutrient-rich. This also makes the area very attractive to fishing operations, and large numbers of dolphins are caught in nets. There is also a significant lack of funding for critically important research in this region.

ABOVE When swimming, river dolphins, such as these Botos, may move their heads in a characteristic fashion, as they scan ahead. Swimming on one side with a flipper trailing in contact with the river bed is an adaptation that helps these dolphins to navigate their special habitat where touch seems to be an important sense.

BELOW River dolphins tend to concentrate in certain parts of river systems, particularly in quiet areas of water and where water channels converge. Here they monitor the prey in nearby fast water currents.

The Marine Dolphins

The ocean-going dolphin species have become a potent environmental symbol in modern society and were also revered by many cultures through the ages. In ancient Greece, where dolphins were strongly linked to the all-powerful gods, killing a dolphin was equivalent to killing a human and was punishable by death. We may not take such a strong view today but the intelligence, gracefulness and typically benign behaviour of dolphins towards humans still have the power to enchant and inspire.

The 34-plus members of the family of oceanic dolphins, Delphinidae (also known as delphinids), seem to be the ocean-travelling nomads of the cetacean world. Typically, they are highly social animals with group sizes ranging from just a few individuals to thousands and, within these groups, they co-ordinate their activities. They include some of the best-known cetaceans: the spectacular Orca, or Killer Whale – the largest of all the dolphins, at 9 metres (30 feet) or more in length – and the ever-popular and adaptable Bottlenose Dolphins.

The names 'porpoise' and 'dolphin' are sometimes used interchangeably, but the delphinids are distinguished from their nearest relatives, the true porpoises (the phocoenids), by a number of anatomical differences; for example, dolphins have cone-like teeth, whilst true porpoises have spade-shaped teeth.

The delphinids are small- to medium-sized cetaceans. They all share the same basic cetacean body-plan (a

torpedo- or spindle-shaped body with prominent flippers and flukes) and the species are visually distinguished using various external features, such as the presence or absence of a dorsal fin and colour patterns. Characteristic differences in their head anatomy reflect differences in lifestyle, particularly prey-specialization. For example, the total number of teeth varies greatly between species. The Bottlenose Dolphin has between 18–26 pairs of peg-like teeth in each jaw, whereas the Orcas have only 10–12 pairs, which are large and conical. Some delphinids have long beaks, others short and some none at all.

ABOVE   Long-beaked Common Dolphins leaping in the Sea of Cortez, Mexico, showing their characteristic colouration and pattern. Like other dolphins, this species tends to be highly social, travels in schools and the mammals co-ordinate much of their behaviour.
OPPOSITE   A fast-moving school of very distinctively marked Hourglass Dolphins, off the South Atlantic island of South Georgia.

These animals occur in a range of colour variations. Like many other marine species, they often have a strong degree of counter-shading, darker on top and paler below. Many species have stripes or streaks and a few are spotted. The dominant colours are black and shades of grey, but some species have splashes of other colours. Recognition of the Common Dolphin species, for example, is often aided by the 'hourglass' pattern on their flanks with its yellow-brown portion towards the head. Colour and patterns may change with age. The Spotted Dolphin species usually become more spotted as they get older and the Risso's Dolphin, which bears prominent white scars, shows its age by the degree of such marks.

Despite their undisputed popularity and a high level of public recognition, the biology of most dolphin species is still only poorly understood. Dolphins are found in most marine waters, although only the Orcas range as far as the cold waters around the polar ice. Some species specialize in living close to shore, others occur far

offshore where their ecological requirements remain a mystery. However, even the offshore species are sometimes seen close to land and some also have populations that continually inhabit coastal waters. As is usual with any attempt to generalize about the cetaceans, there are also some contrary 'marine dolphins' that can be found in rivers. For example, parts of the populations of the Tucuxi (the smallest dolphin) and the Irrawaddy Dolphin are regularly found far upstream, and this is clearly still within their natural ranges.

Some dolphin species favour upwelling areas, which are usually rich in nutrients and prey, because the water circulation brings these up from the depths. The seemingly nomadic nature of many marine dolphins reflects the fact that they travel from one resource-rich area to another. Whilst these 'resources' are most commonly prey, they may also be seeking other things, including more favourable water temperatures.

Foraging strategies vary between species and sometimes between populations of the same species, and many show a high degree of adaptability. Prey, for most dolphin species, means fish or cephalopods (octopuses and squid) but a few, most famously some Orca populations, prey on other marine mammals. Novel feeding strategies can develop within individual populations and are learnt by successive generations.

On a global scale, very few dolphin species or subspecies are currently regarded as threatened or endangered. However, the available data for many dolphin populations are poor or has not been fully evaluated, and problems in determining where discrete populations begin and end also make conservation assessments difficult.

ABOVE   A Tucuxi – a small but robust dolphin found in both salt and fresh waters on the Atlantic coasts of South America. This one was photographed off Brazil.
OPPOSITE   Atlantic Spotted Dolphins off Little Bahama Banks in the Atlantic. The darker, spotted animal exhibits typical, very distinctive adult patterning. Younger members of this species are typically born without spots and these develop over time but there is also considerable variation in the degree of spottiness between adult individuals.
BELOW   An old, white-faced and heavily scarred Risso's Dolphin seen off San Diego, California.

One subspecies of the Long-finned Pilot Whale, previously found in the North Pacific, is thought to be extinct already.

Hector's Dolphin is only found in coastal waters of New Zealand and there are as few as 7,400 remaining, mainly around the South Island. The species is divided into three genetically distinct groups. The smallest, the North Island group – classified as critically endangered and recently renamed Maui's Dolphin – numbers less than 100. Inshore waters have become particularly dangerous places for these marine mammals. Fisheries operations abound and boat traffic, including fast-moving leisure craft, is increasing. Young Hector's Dolphins are sometimes struck and killed by boats.

ABOVE A small group of rare Hector's Dolphins off Kaikoura in New Zealand exhibit their distinctive black and white patterning and rounded dorsal fins. New Zealand takes conservation very seriously and has created sanctuary areas (where fishing may be controlled), sponsored new research and made protective laws. Time will tell if this will be enough to save the Hector's and Maui's Dolphins.

BOTTLENOSE DOLPHINS

These are the best known of all of the cetaceans due to their inshore habitat (although they can be found offshore too) and robust nature. They are medium-to-large dolphins, with distinctive beaks, and are usually slate grey or charcoal in colour, with a paler ventral surface. Until recently the 'Bottlenose Dolphin' was regarded as the world's most cosmopolitan dolphin. However, the populations have now been 'split' into two species: the Indo-Pacific Bottlenose Dolphin (*Tursiops aduncus*) was awarded specific status species; while all the other populations (at least for the moment) are known as Common Bottlenose Dolphin (*Tursiops truncatus*). Bearing

in mind that this latter species, far from being numerous, may actually be uncommon or even rare in some areas, the use of this name is perhaps unfortunate.

The female Indo-Pacific Bottlenose Dolphins of Shark Bay in Australia have been observed holding sponges on the ends of their beaks while they dabble in the seabed sediments. They are looking for fish, but the sponges appear to protect their sensitive snouts from the spiny creatures that also live down in the sand and mud. They are, therefore, the first wild cetaceans to be recorded using a tool.

Female Bottlenose Dolphins can live for more than 50 years, and males for 40; the females start to breed when they are in their early teens, or even a few years earlier, and males mature at similar ages. Fully-grown males are typically bigger than females, and sizes may

BELOW   Indo-Pacific Bottlenose Dolphins feeding on a 'baitball' of sardines in the Indian Ocean off Transkei, South Africa.

vary significantly between populations. A Florida Common Bottlenose Dolphin might be about 2.5 metres (over 8 feet) long when fully grown, but an adult British Bottlenose could be more than 1 metre (3 feet) longer than this, possibly an adaptation to living in a colder environment.

ABOVE Its extra rotundity reveals that this Common Bottlenose Dolphin, from the Little Bahama Banks in the Atlantic, is pregnant. Unlike most other mammals, when the calf is born, it emerges tail first – an adaptation to living in water.
OPPOSITE While sizes and colour patterns may vary, one thing that all Bottlenoses seem to have in common is their 'fission–fusion' style of behaviour, meaning that individuals often associate together in small groups but that the composition of these groups changes over time. This particular group of Bottlenoses was photographed off Japan; in the foreground is a mother and calf.

Pregnancy lasts for 12 months and, like all other cetaceans, a single calf is usually produced. The new-born enters the marine world tail first – unlike most terrestrial mammals which tend to be born head first – and one of the mother's first acts – or that of an 'auntie' or another assisting group member – is to help it up to the surface to take its first breath. Calves remain with their mothers for several years and may continue to take milk up to the age of five years or more. This long period of dependency allows the young dolphin to learn how to be a fully functioning member of its society. There is some evidence that breeding is a seasonal occurrence but, like other dolphin species, the correlation between a season and breeding seems generally to be rather weak.

Some of the best-known individual Bottlenose Dolphins, such as the animal known as Fungie, which lives in Dingle Bay in the west of Ireland, are famously solitary. However,

group living is the usual rule and the dolphins living further offshore typically form larger groups. Mother and calf pairs always stay close together, the young calf typically swimming under the mother's tail, close to the milk supply, in a characteristic 'following position'. Other individuals are also often seen in close proximity to each other, sometimes almost touching, and frequently echoing each other's movements through the water. These closely associated animals may be males and may be playing or co-ordinating their hunting of either prey or receptive females for mating.

Bottlenose Dolphins eat fish and cephalopods, but their diets can be highly varied. They seem to have few natural predators, except for some sharks, and dolphin groups may act together to drive them away or kill them. From eyewitness reports and from some of the severe bite marks seen on the bodies of dolphins, it is clear that the dolphins do not always win. It is not clear if mammal-eating Orcas ever attack and eat Bottlenose Dolphins.

OPPOSITE Orcas are highly adaptable and efficient predators. The Orca seen in this extraordinary photograph, which was taken off New Zealand, was part of a group that later killed and ate the Mako Shark also pictured.

BELOW Common Bottlenose Dolphins leaping spectacularly in synchrony off the Bahamas. More usually, only their slate-grey backs are seen at the surface. However, with care, individual dolphins can be recognized based on the scars and nicks on their backs and dorsal fins. Scientists usually use photographs to confirm the identification of individuals and the science of 'photo-identification' is now a primary method in international cetacean research.

ORCA

Another species about which we know a fair amount – at least for some populations – is the Orca, or Killer Whale. The Orca's social organization is probably the best described of any cetacean and there is such diversity within this taxon that it is no longer appropriate to think of them as a single species.

Orcas are the largest member of the dolphin family and the most superbly adapted marine predators of all. The old species name – Killer Whale – reflects the fact that some populations specialize in feeding on other marine mammals, including other cetaceans. So, these animals became the 'whale-killers' or 'Killer Whales'. However, not all Orcas eat whales or other mammals; this is one reason why the name Orca – from their scientific name *Orcinus orca* – is preferable and is being used increasingly.

Orcas have a striking black-and-white colouration, including a large white eye patch. This pattern may help these mammals, which have good eyesight, to monitor each other's movements above and below the water. Adult male Orcas may weigh twice as much as the females – about 6,000 kilograms (13,230 pounds). The males also have proportionally larger flippers and tail flukes and their most obvious characteristic is the remarkably tall dorsal fin.

Evidence from genetic studies worldwide shows that Orca populations have diverged, primarily as a result of feeding specialization. For example, there are distinct

populations of mammal-eating and fish-eating Orcas off the west coast of the US and Canada that should probably be considered as two separate species.

The main division is between 'residents' and 'transients'. The residents are a remarkably stable population. During studies spanning a quarter of a century, no individuals appear to have left the population (other than those that have died) or joined it (other than by birth). These resident Orcas feed only on fish.

OPPOSITE   Orcas in Patagonia have even learnt to catch their prey while it is still on shore. Here a group of juvenile Orcas challenge a large male Southern Elephant Seal.
OPPOSITE BELOW   An Orca spectacularly breaching off the San Juan Islands in the East Pacific.
BELOW   Dramatic and synchronous surface swimming by a pod of Orcas in Norwegian waters.

Transients, however, have quite a different lifestyle, and specialize in feeding on marine mammals. Orcas live in social groups called pods and there is much more movement between the typically small pods within the transient population than between the larger pods of residents. The latter are also more predictably found in certain locations than the transients.

While there are several ongoing Orca research programmes around the world, there are also many other Orca populations that have not been well studied. Given that these are highly adaptable and intelligent animals, it may be dangerous to try to extrapolate what we understand from one population to another. Clearly, however, they are all capable of complex behaviour. For example, Orcas off Norway have been seen to collaborate closely to herd herring from the deep water towards the surface, and even use blasts of bubbles and noise to help drive the fish into a tight ball, on which they can then efficiently feed.

The beaked whales

Beaked whales are, quite simply, fascinating! They form the least well-known group of cetaceans and, in fact, some of them are more like mythical creatures than real animals, with sightings of some species rarer than those of Scotland's Loch Ness Monster.

Longman's Beaked Whale is a good example: prior to 2003, only two individuals had been described from skull parts cast up on the shore in Australia and Somalia. The first time that the external appearance of this beaked whale could actually be recorded was from four strandings in the Indian Ocean in 2003. This discovery also significantly extended the known range of this species.

Until 2010, when two specimens washed up on a New Zealand beach, the Spade-toothed Beaked Whale was only known only from the skull parts of three individuals found in Chile and New Zealand. New species of beaked whales are still being classified on a regular basis. For example, Perrin's Beaked Whale was described as recently as 2002 (and named by its discoverers after the distinguished cetologist, Bill Perrin).

ABOVE A Cuvier's Beaked Whale – its distinctive head shape has also caused it to be known as the Goose-beaked Whale – off Cape Hatteras in North Carolina. Note the two small teeth protruding characteristically from the lower jaw.
OPPOSITE The heavily scarred backs of a school of Baird's Beaked Whales seen off Baja California.

The beaked whales (or ziphiids) range in length from 3–13 metres (10–43 feet), the largest probably being Baird's Beaked Whale. Most species have a surprising appearance, including rather small flippers, and beaks that merge with the rest of the head. Another characteristic is a deep, forward-pointing, 'V'-shaped fold in the skin in the throat region. Most startling of all is their dentition. With the exception of Shepherd's Beaked Whale (which has many teeth), the ziphiids have no teeth or just a few vestigial ones in the upper jaw and one or two pairs below, which are only normally seen in the males. In some species, the teeth extend outside the mouth like tusks and, in the aptly named Strap-toothed Beaked Whale, they wrap around the upper jaw, actually limiting the ability of the mouth to open. It is assumed from scarring that the males use their large teeth to spar with each other, although this has never been witnessed.

Our relatively poor knowledge of beaked whales can be largely explained by their biology. All beaked whales are usually found offshore and dive deep to find their prey of deep-dwelling squid or fish. Some, at least, also seem to avoid boats. Their behaviour on the surface appears to be rather simple, but there may be far more sophisticated behaviour and communication going on deep underwater where (for the moment) we cannot follow. However, there is some social cohesion because after their long dives, which last 20–40 minutes or more, animals in groups usually surface close to each other.

None of the beaked whales are currently classified as endangered, but there is a considerable lack of information. The king of the ziphiids, Baird's Beaked Whale, is only found in the North Pacific and continues to be hunted in Japan, which sets its own quota, claiming that the International Whaling Commission has no jurisdiction over this 'small cetacean'. There is growing international concern about this hunt, including the methods used to kill such a large animal.

ABOVE The extraordinary Blainville's Beaked Whale, which has two large teeth that erupt from its strongly arched lower jaw. The teeth can be seen in this photograph protruding like horns above the animal's head. Its relatively small head and flippers are features shared by all the beaked whales.

LEFT This breaching Blainville's Beaked Whale has barnacles attached to its protruding teeth. It was photographed in Hawaii.

OPPOSITE The huge open mouth of a Humpback Whale as it engulfs a vast mouthful of water and prey while lunge-feeding in the cold and productive waters off Alaska.

The 'great whales'

The term 'great whales' is usually used to describe the 14 mainly large, filter-feeding baleen whale species (also known as the mysticetes) and the Cachalot (or Sperm Whale). The 14 baleen whale species are divided into four families. The rorqual whales, belonging to the family Balaenopteridae, are all very similar in appearance: long, thin (with the exception of the Humpback Whale) and fast moving. These whales are said to 'gulp' their prey. The other three families are Balaenidae (the Bowhead and Right Whales) and Neobalaenidae (Pygmy Right Whale), which are 'grazers' (or 'skim-feeders') and, finally, Eschrichtiidae, which contains only the Grey Whale, a species that 'siphons' up its prey.

The Great Gulpers

These are the greyhounds of the whale world and vary greatly in size. The Minke Whale, which was recently divided into two species – Common and Antarctic – is relatively small with maximum lengths of 9.2 and 10.7 metres (31 and 35 feet) respectively. Next in size is the Bryde's Whale, which reaches 15.5 metres (51 feet);

followed by the Humpback Whale, which reaches 16 metres (53 feet), the Sei Whale, which reaches 20 metres (61 feet), the enormous Fin Whale, which reaches 27 metres (89 feet) and the Blue Whale, which reaches about 34 metres (112 feet).

The genetics of these rorqual whale species are currently causing considerable interest. The dwarf form of the Minke Whale, found in Antarctic waters, may be declared a third separate species. Bryde's Whale is also likely to be reclassified as more than one species. The Blue Whale also has a smaller form – a mere 24.4 metres (80 feet long) – and this is presently treated as a subspecies known as the Pygmy Blue Whale.

Rorqual whales are basically highly streamlined and fast-moving animals with relatively short baleen plates and many throat grooves. These grooves allow their throats to balloon out when they feed and 'gulp' huge mouth- and throatfuls of sea water and prey. The tongue (the size of an elephant in the Blue Whale) and contraction of the throat grooves force the water out through the filtering baleen plates which collect the prey.

MINKE WHALES

Like the other rorquals, Common and Antarctic Minke Whales are strongly counter-shaded, dark above and pale below, but the Common (including its Dwarf subspecies) also has a characteristic well-defined white flipper mark. Both species of Minke take a variety of prey, which seems to vary with the season and includes various fish species as well as planktonic crustaceans. Minke Whales are sometimes seen in large groups of several hundred animals, but are more often seen alone or in small groups of just a few individuals. The species exhibit some migratory behaviour and populations divide up according to sex or age. Both species have similar life histories and – unlike most other whales – breeding may occur every year. Calves are born after a pregnancy of about ten months and are a little under 3 metres (10 feet) in length. The young feed on milk for at least four to five months and become sexually mature when six to eight years old. Minke Whales are quite vocal and produce a range of noises, which presumably help them to co-ordinate their activities.

These whales are currently hunted in the North Pacific and in Antarctica by Japan, and in the North Atlantic by Norway and Iceland. Because they are relatively small, Minke Whales were ignored during the heyday of industrialized whaling but, as the larger rorqual whales were depleted and they (along with other smaller species) were still relatively plentiful, attention increasingly turned to them.

Another small rorqual – Omura's Whale – has recently been described as new to science. It inhabits the western Pacific and little is currently known about it.

RIGHT   A Dwarf Common Minke Whale photographed on the Great Barrier Reef in Queensland, Australia.
BELOW   An Antarctic Minke Whale in the Southern Ocean Sanctuary.

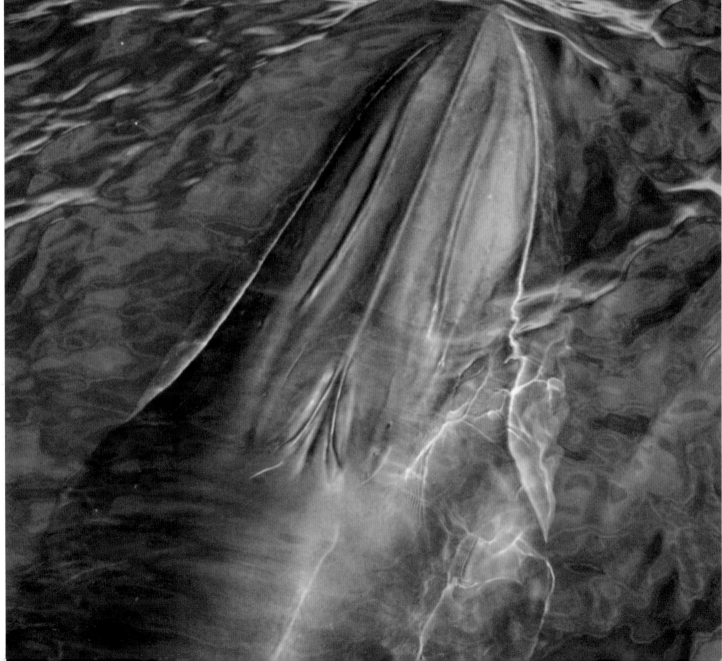

BRYDE'S WHALE

This is the least well-known of the baleen whales. It has three distinct ridges on the top of its head, which help to distinguish it from the strikingly similar Sei Whale, which only has one. This species is seen in tropical and warmer temperate waters year-round, although there is a general movement in the winter towards the equator. This species is also usually seen alone or in small groups. Its breaching activity is sometimes spectacular. Occasionally, a single individual may breach many times without pause, leaving the water over and over again in an almost vertical orientation. This species produces powerful and characteristic low-frequency sounds. Breeding grounds are poorly defined, but pregnancy appears to last for about 11 months. At birth the young are about 4 metres (13 feet) long, and feed on milk for about six months.

Bryde's Whale were one of the last species to be commercially hunted due to their relatively small size.

SEI WHALE

This is a very similar species and undertakes a much more distinct migration, typically feeding in the summer in the polar regions and returning to lower latitudes to breed in the winter. Pregnancy lasts about a year. The new-born calf is 4.5 metres (almost 15 feet) in length and weaned at seven months, at which time it will have already completed its first migration and will be with its mother in the cold feeding grounds. The migratory nature of the Sei Whale and the fineness of its baleen relate to the preferred prey of this whale species – small planktonic copepods. They appear to feed mainly at dawn, skimming through patches of plankton. Fish and cephalopods are also sometimes reported in their diet.

The Sei Whale becomes sexually mature when about ten years old, although the age of maturity declined by two to three years for some populations significantly impacted by whaling. Sei Whale populations are now thought to be recovering from the effects of whaling conducted in the late 20th century.

OPPOSITE ABOVE   Because of its distribution, the Bryde's Whale is also sometimes known as the Tropical Whale. It has been reported from all tropical and temperate waters in the North and South Pacific, the Indian Ocean and usually between 40°N and 40°S.

OPPOSITE BELOW   An excellent view of the top of the head of a Bryde's Whale, showing the three ridges above the upper jaw that help to identify these whales and also its paired blowholes, which are tightly closed.

ABOVE   These Sei Whales – an adult and calf – photographed off the Azores Islands in the North Atlantic, show the streamlined bodies typical of all the fast-swimming rorqual whales. Several of the rorqual whales can be easily confused with each other, especially at a distance. Sei Whales have a single head ridge and behaviour and distribution can also help to distinguish between species.

HUMPBACK WHALE

Found in all the world's oceans, the Humpback Whale is a truly magnificent animal, the acrobat of the great whale species and a fine singer. Its spectacular displays of breaching and tail or flipper slapping mark it out as a very energetic animal. It can be easily recognized by its exceptionally long flippers (about one-third of the body length) and the characteristic bumps on its head. The patterning and shape of the underside of the tail have allowed thousands of individuals to be followed over many years. The females are usually a little larger than the males and the calves are 4–5 metres (13–16 feet) long when born, following a pregnancy of about 11 months. Calves are weaned when they are about a year old and become sexually mature at about five. Most Humpback Whale populations undertake regular and extensive migrations.

Humpback Whale society features both long- and short-term associations, but they are most famous for the remarkably melodious and complicated songs produced by the males in winter. Remarkably, all the males in a population sing the same song, although it changes over time. The songs appear to be used in courtship, so some presumably sing in a more attractive way than others. The males also fight for the females, and this may help to explain the purpose of the big solid lump under their chins.

Commercial whaling decimated some Humpback Whale populations, but a number are now recovering. The most immediate threat is probably accidental capture in fishing nets.

LEFT   Humpback Whales – like this animal photographed off Hawaii – are the most acrobatic of the great whale species, often making tremendous leaps above the sea and frequently even back-flips as seen here. Note the distinctively large flippers of this species, the lumpy head and the prominent throat groves.

FIN WHALE

Apart from its huge size and shape, recognition of the Fin Whale is helped by the fact that it has an asymmetrically coloured head. The underside of the right-hand side – and part of the baleen on the same side – is white (as shown below), whereas the entire left side is dark. Asymmetrical colouration is rare and may reflect the fact that the animal typically feeds underwater with its left side facing up.

Female Fin Whales are a little larger than males and become sexually mature when seven to eight years old. The males mature a year or so earlier. A new-born calf is already 6–7 metres (20–23 feet) long and weighs some 2 tons. Weaning occurs after six to seven months.

The distinct north–south migration of Fin Whales in the Southern Hemisphere is not well echoed by the Northern Hemisphere populations, where wintering grounds have not been identified. It may be that Fin Whales in the North Atlantic can find all of their needs over a smaller range and without always making such long migrations.

A large Fin Whale represented a good haul for whalers and tens of thousands were taken before the commercial whaling moratorium came into effect in 1986. Populations are now said to be either recovering or stable. Entanglement in fishing gear is a problem, and even the great size of these whales does not protect them from being struck by the increasingly swift vessels now criss-crossing the world's oceans.

BLUE WHALE

The gigantic Blue Whale is, conveniently, blue on top – at least in certain lights. Mottled patches of grey that can shine turquoise or even silver on a sunny day cover the skin on their backs. Underneath, Blue Whales can be white, blue or yellow (hence one of their other names, the Sulphur-bottom Whale) and the yellow colour is caused by a hitchhiking layer of microscopic plants, called diatoms. The whales have 55–88 deep throat grooves (shown in the picture above) and their blow, when they exhale at the surface, is very impressive; on a still day this can be over 10 metres (33 feet) high.

This species is found in all oceans and typically migrates annually into polar waters in search of its primary prey, Krill, which are planktonic crustaceans reaching just a few centimetres in length. The Southern Hemisphere Blue Whales are a little larger than their northern relatives; all Blue Whales reach sexual maturity at about eight to

ABOVE A Blue Whale feeding on plankton. It has just scooped up a throat and mouthful of water and prey.

OPPOSITE Their huge size and the tall column of spray that they produce at the surface when they exhale, helps Fin Whale recognition at sea. This individual, from the Sea of Cortez in Mexico, also exhibits the asymmetrical white coloration of the right side of its head (the other side being dark). A fast-swimming rorqual whale, the Fin Whale's typical swimming speed is 5–8 knots, although it can reach up to 15 knots for short bursts.

ten years and may live for up to 90 years. Pregnancy lasts up to a year; the new-born is 6–7 metres (20–23 feet) long and weighs about 3 tons. Infants are weaned when they are six to eight months old, by which time they have almost doubled in size.

The largest animal is also the loudest. The moans made by Blue and Fin Whales are low-frequency tones, below the range of human hearing, and loud (185–190 decibels). They can probably be heard for many thousands of kilometres, meaning that both these species may be able to communicate across entire ocean basins.

Natural mortality of Blue Whales is rarely reported and their principal predator is the Orca, as evidenced by the scars seen on some animals. With the advent of fast-moving and wide-ranging factory-whaling vessels, the Blue Whales became the whaler's most prized target because of their great size. Today they are still endangered in Antarctica, where only a few hundred remain, but there are reported signs of an increase in the Northern Hemisphere population from Iceland and California. Because of their dependence on seasonal plankton blooms in the Arctic and Antarctic, they may be especially vulnerable to climate changes affecting polar waters.

BELOW   A Blue Whale beginning a dive off Mexico.

The Great Grazers

The name 'right whale' derives from the conclusion of the early whalers that they were the 'right whales' to hunt because they were large and slow, easy to approach and, when harpooned, conveniently floated. They could also be found close to shore.

These are the grazing whales, robust animals with a wide girth, big heads (about one-third of their body length) and strongly arched jaws that support very fine baleen filters. They swim through plankton patches with their mouths open, filtering out the plankton, particularly copepods, which are tiny crustaceans just a few millimetres long. There are four Right Whale species – North Atlantic, North Pacific, Southern and Pygmy – although Pygmy has quite a different biology and has been placed in a different family. Closely related to the larger Right Whales is the Bowhead Whale, which shares some of their characteristics. All these whales, with the exception of the Pygmy, lack a dorsal fin.

RIGHT WHALES

Maximum lengths for the Right Whales are about 18 metres (60 feet), and the newborns measure 5–6 metres (16–20 feet). The males are mature at a length of about 15 metres

BELOW   A Southern Right Whale breaching off Argentina.

(50 feet) and the females at about 0.5 metre (18 inches) more when they are about ten years old. These bulky animals have head armour, large concrete-like growths known as callosities that may be used in defence and also by the males when competing for access to the females. Rather surprisingly for seasonally breeding animals – calves are normally born in the middle of winter – mating in Right Whales goes on year-round and, to support this activity, the males can claim to have the largest testes in the animal kingdom, each weighing about 1 ton. When not breeding, Right Whales are usually seen on their own or in small groups. Sometimes they are seen to feed side-by-side, although the significance of this is not clear.

On a clear, still day, the blow of Right Whales sprays up from their double blowhole to produce a distinctive 'V' shape, as does that of the Bowhead. Both the Southern and both species of Northern Rights make migrations between feeding and breeding grounds and tend to stay close to land as they travel. Breaching seems to be a common activity in the Southern species, which often makes a series of mighty leaps from the water, one after the other.

The North Atlantic and North Pacific Right Whales were the earliest targets of organized whaling. Currently the population of North Atlantic Right Whales off the east coast of North America numbers only around 300–350 individuals. On the other side of the Atlantic the species is either very rare or extinct. There was a sighting in the mid-1990s and, more recently, another to the west of Iceland, but these could be individuals wandering across from the North American population. Similarly, the North Pacific Right Whale population is small and scattered and, in the west, just a few hundred can be found in the summer in the Sea of Okhotsk.

The main threats to the two northern species now seem to be ship strikes and entanglement, and more than half of the surviving animals in the North Atlantic population have been struck or entangled already. There is better news in the Southern Hemisphere for the Southern Right Whales, which now number some 7,000, and where populations are showing signs of recovery.

The Pygmy Right Whale is the smallest baleen whale, reaching about 6.5 metres (21 feet) and occurs in the temperate and subantarctic parts of the Southern Ocean. Little is known about its biology and there have been few observations of living animals. Its fine baleen indicates that it specializes in copepods and it resembles a cross between a Right Whale and a Minke Whale with which it is often confused. It is a rather sleeker animal than the other Right Whales and also lacks their characteristic callosities.

BOWHEAD WHALE

The final large, grazing whale spends its entire life cycle in Arctic waters. These whales became famous when research indicated that some were over 200 years old. It is another slow-swimming and robustly built whale, reaching some 20 metres (65 feet) in length. It can dive longer than other baleen whales, sometimes staying under for almost an hour, and is able to break thick ice with its massive head, although it does not have any callosities. Pregnancy may last over a year (maybe as long as 14 months) and the 4.5-metre (15-foot) long new-born is weaned before it is a year old. Sexual maturity doesn't occur until the whales are 15–20 years of age. This

lengthy 'adolescence' provides a long learning period, which could be invaluable in helping young whales survive in their extreme habitat.

Bowheads feed primarily on planktonic crustaceans and, like the Right Whales, 'skim-feed' with their mouths open. However, stomach contents from animals taken in hunts show that they also feed on other prey and appear quite adaptable. Their behaviour is similar to that of the Right Whales and they are usually seen as solitary animals or in small groups.

After the Right Whales were hunted close to extinction in the north, the Bowheads became the next targets for early commercial whalers and their populations were successively decimated. There are currently about 8,000 in the western Arctic region and a few hundred in the Davis Strait–Baffin Bay, Hudson Bay–Foxe Basin and Okhotsk Sea areas. Fewer

OPPOSITE   The top of the head of an unusually pale-coloured Southern Right Whale calf in Argentina. Its colour will probably darken as it grows.
ABOVE   A North Atlantic Right Whale – one of the most endangered species – looms up out of the cold waters of the Bay of Fundy in Canada. Note the strongly arched jaws.

than 100 are left in the Svalbard–Barents Sea region and this population is regarded as critically endangered. The other small populations are all either regarded as endangered or vulnerable. The western stock is still hunted by indigenous peoples from Alaska, western Canada and Chukotka (a Russian province), and the Davis Strait–Baffin Bay population is hunted by the Inuit of eastern Canada. There is some proof that the western Arctic region population is recovering slowly, but this evidence is lacking for the other populations.

ABOVE The tail flukes of the great and long-lived whale of the Arctic, the Bowhead Whale, at the ice-edge in the Canadian Arctic.
LEFT A Bowhead Whale with a strand of seaweed across its snout surfacing near to the ice-edge. New scientific techniques have recently proved that animals of this species can live for hundreds of years. One was recently found to be over 200 years old.
OPPOSITE Head-on: the pointed snout of a curious Grey Whale calf, revealing the sensory hairs on its upper surface, San Ignacio Lagoon, Mexico.

The Great Bottom-feeder

The final great whale, the Grey Whale, is the only whale that specializes in feeding on the seabed. It is a shabby, grey, barnacle-encrusted species, which reaches a length of about 15 metres (50 feet). In body shape it is somewhere between the streamlined rorquals and the stocky Right Whales. It makes prodigious migrations and is a species once greatly feared by the whalers because of its 'ferocity'. This reputation originates mainly from the days of the inshore whalers, when some Grey Whales would turn back and attack the boats used to harpoon them. They are often encountered in small groups and some individuals have been seen to regularly accompany each other.

GREY WHALE

This species is now only found in the North Pacific, although it was present in the North Atlantic as recently as the 18th century. It is another distinctly coastal species, now divided into only two populations on either side of the ocean. The western population can be found from Korea in the south to the Okhotsk Sea in the north. The eastern population ranges from Mexico, at the southern end of its migration, to the Bering, Chukchi and Beaufort Seas at the northern end. The well-studied annual migration of the Californian population is, at maximum, a 20,000-kilometre (12,420-mile) round trip.

Grey Whales are about 4.5 metres (15 feet) long when

born, and become sexually mature when some eight years old. Their life cycle and migrations are strongly entwined. In the case of the Californian population, all the calves are born into the warm and shallow waters of the lagoon system of Baja California. The young whales are weaned when about seven months old. Pregnancy is notably long – some 131/2 months – and sexual maturity dawns when the whales are 5–11 years old.

Coming to the surface to breathe, they are often besmirched with mud from feeding on the seabed where they orientate themselves onto one side (usually the right, judging from the wear patterns), part bury their heads and suck up invertebrate-rich sediment. Suction is provided by retraction of the tongue combined with the expansion of their three to seven throat grooves. The mixture of mud, food and water is then pushed through the baleen plates by the tongue. Food (and other things) are retained and swallowed. They may also stir up the sediments first before ingesting the resulting water-and-mud cocktail and spit out things that they don't like, although stomach contents reveal that many non-edible articles get ingested, too. Grey Whales can also feed on

organisms in the water column and may even sit in strong currents with their mouths open letting their baleen plates 'net' passing food. Thus, it could be said that baleen whales invented fishing before we did: the Right Whales 'trawl' for their prey and the Grey Whales dredge or net.

There are grave concerns about the western Pacific population of Grey Whales – which may only number 100 animals – particularly following the opening up of their principal feeding ground for oil and gas development. Grey Whales are also vulnerable to ship strikes and bycatch. The eastern stock is hunted in Russia but has been steadily increasing, although a high level of mortality (observed as strandings along its well-monitored migration route) has been reported in recent years. The cause of these strandings has been hotly debated, including the notion that the population had reached the maximum size that could be naturally supported.

OPPOSITE The 'great bottom-feeder' – a juvenile Grey Whale explores the giant fronds of a Californian kelp forest.
ABOVE A sub-adult Grey Whale surfaces in the tranquil waters off Isla Salsipuedes, Mexico.

The others

CACHALOT (SPERM WHALE)

The following species do not fit very neatly into any of the groups already described. The biggest 'misfit' is the Sperm Whale. While, in terms of size, it is undoubtedly a great whale, the Sperm Whale, unlike the other great whales, has teeth. Their bizarre common name originated from the milky liquid 'wax' found in their heads, which was originally mistaken for semen. The perpetuation of this mistake is inappropriate and the alternative common name, Cachalot – derived from the common name for this species used in French and other languages – is used throughout this text.

Because Cachalots have teeth, they have long been classified as part of the toothed whale grouping. However, there are suggestions that, really, they are more closely aligned with the baleen whales. Whatever their origins, they are singular in appearance, being dark brown or black in colour, with a big square-ended head that is a third of the body length in adult males. The males, which are significantly larger than the females, can reach maximum sizes of 20 metres (65 feet) and weigh over 40 tons. They can dive very deep and specialize in feeding on squid.

ABOVE   A Cachalot or Sperm Whale comes to the surface to breathe. Note the white insides of its mouth.
BELOW   A very young calf in a group of Cachalots still with bands around its body that formed when it was curled up in the womb.

ABOVE The small dorsal fin and knobbly and arched spine of a Cachalot as it dives.

RIGHT The distinctive exhalation (or blow) of a Cachalot, which is directed towards the left side of its body.

The lower jaw looks disproportionately small and contains 50 or so conical teeth, which can weigh up to 1 kilogram (over 2 pounds) each. There are usually no teeth apparent in the upper jaw. The skin towards the rear of the whale is highly wrinkled and they have a poorly defined dorsal fin or hump.

These huge animals can produce a 5-metre (16-foot) blow, which projects forwards and to the left from their curiously asymmetrically placed blowhole. The basic biology of this species is also curious. Males and females live in different parts of the planet for most of their lives. Only the large males can be found in the colder latitudes (beyond about 45 degrees) and up to the ice-edge in both hemispheres. Females are extremely rare in these colder waters.

The sexes meet when the mature males migrate into the tropical waters to breed. Both sexes roam widely within oceans and the females live within stable social groups. The new-born Cachalot is some 4 metres (13 feet)

long and is produced after a very long, 16–17-month pregnancy. Weaning does not occur until well after the young whale is able to catch solid food and animals as old as 12 years have been seen still suckling. Females are sexually mature at 8–12 years and males are physically and sexually mature at about ten years but, under normal circumstances, may not breed until they are in their twenties.

The female 'family groups' consist of closely-related individuals and calves are looked after collaboratively.

Males leave the family when they are about six years old and make their way to the cold waters where they may join bachelor herds. Here they stay until they are big enough to be able to return and mate successfully with the females.

Mature males will range between female groups searching for receptive partners. Historically, whalers focused on the larger males, primarily seeking the greatest harvest of highly valuable oil. One knock-on effect of the ongoing lack of mature males has been unexpected: females will not accept the younger males as mates and

this means fewer calves are probably being produced.

There are two other species that bear the name Sperm Whale: the Pygmy and the Dwarf. They are not miniature versions of their namesake (and are in a different family) but are robust and distinctive little whales. The Pygmy reaches about 3.3 metres (11 feet) long, and the Dwarf is a little smaller. Their heads are bulbous and they have short, narrow mouths with 12–16 (Pygmy) or 7–12 (Dwarf) curved, needle-shaped teeth in each lower jaw. Their biology remains little known, but they are mainly found in deeper waters and along the continental shelves and seem to feed at depths of 200–250 metres (655–820 feet), mainly on cephalopods. Both species have been subject to a small degree of hunting in the past, but gill nets and ingestion of plastic debris seem to be their most pressing current problems.

BELOW   One of the ocean giants: a Cachalot seen in its breeding grounds off the Azores Islands in the Atlantic.

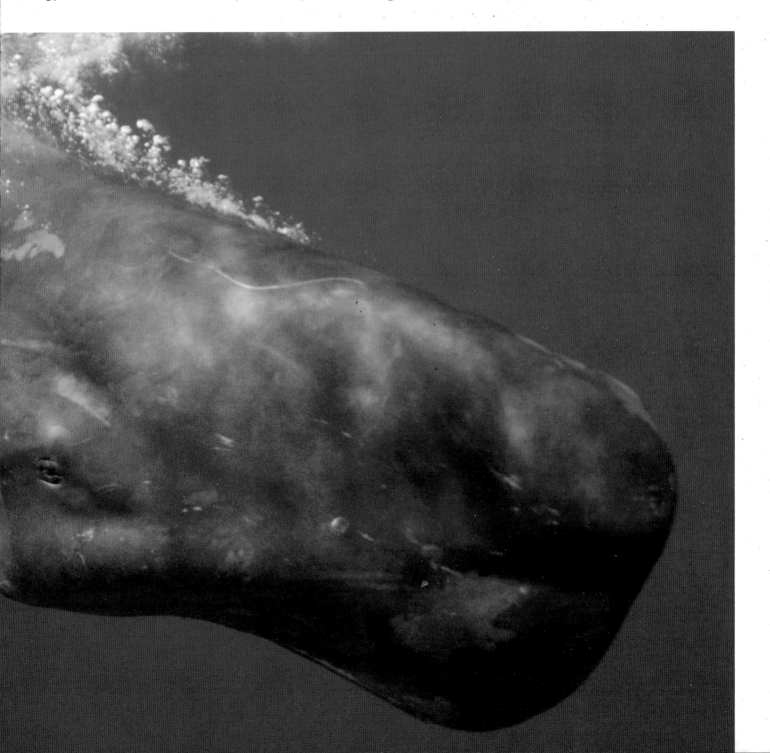

WHITE WHALES

The two closely related 'white whales' are the Beluga and the Narwhal. Their flexible necks mean that these whales can turn their heads and look at you, a rare ability in cetacean circles. Another unusual feature is the Beluga's ability to change the shape of its head, giving it a range of 'expressions' which, in reality, are probably an adaptation to feeding, for example, allowing it to suck up prey, and echolocation – changing the shape of the acoustic 'lens' in its head. They both live year-round in cold circumarctic waters. The Narwhal, in particular, is rarely seen far from the ice-edge and makes seasonal movements with the ice as it expands in winter and contracts in summer.

At birth, a Narwhal is about 1.6 metres (5 feet) long. The calves are uniform grey or browny-grey and this changes over the years to a flecked pattern of dark brown or black on the adults' backs. Adult Beluga Whales are usually a pure, snowy white, while their new-born calves are grey. Male Belugas can reach a maximum size of over 5 metres (16 feet); the females are a little smaller, reaching at most a little over 4 metres (13 feet). Females become sexually mature when they are about five years old and males a few years later. They mate in the late winter or the spring and the calves arrive some 14–15 months later in the summer. The life cycle of the Narwhal is very similar.

Living in an environment where the temperature is rarely much above freezing obviously calls for a number of special adaptations. Calves are wrapped in a thick blanket of blubber, and weaning does not occur until late in their second year. Calves are typically cared for within a group of females. Beluga Whales are highly social and are most often seen in groups of about 15, but occasionally numbering in the thousands. Their behaviour is described as often playful and noisy.

Narwhals continue to be heavily hunted in Greenland and the eastern Canadian Arctic, and Beluga Whales are similarly threatened. In addition, the Beluga population of the St Lawrence Estuary has become infamous because of its highly polluted condition and associated disorders, including cancers and hermaphrodism (having the organs of both sexes). Increasing vessel traffic in Beluga habitats is another growing concern.

OPPOSITE   A Beluga Whale, surrounded by many others, uses its flexible neck to peer around above the water surface in the Canadian Arctic.

BELOW   The unicorns of the sea: a group of Narwhals in the Canadian Arctic showing their remarkable long spiral tusks, which are actually modified teeth, projecting from the front of the head. These tusks are unique to this species.

Chapter Two

LIFE HISTORIES AND STRATEGIES

Out in the open ocean there is nowhere for cetaceans to hide – except behind each other. Prey tends to be concentrated where the marine conditions are favourable, but these sites are not necessarily permanently fixed. So, finding adequate prey demands effective navigational and hunting skills, as well as energy-efficient locomotion. For some cetaceans, prey can be located half the planet away from their breeding grounds.

The marine environment is also colder than body temperature, posing a real challenge for mammal species, and light rapidly disappears with depth, requiring different sensory capabilities to sight, which most terrestrial mammals heavily rely on. Cetaceans have evolved many different strategies to allow them to survive in an environment that most other mammals, including humans, would find entirely alien. In this chapter we consider the inner working of cetaceans and how their adaptations, coupled with their intelligence and behaviour, allow them to survive.

Basic biology

WHY DO DOLPHINS AND WHALES RESEMBLE FISH?

Cetaceans are streamlined and use their large tails to propel themselves through the water so, in these respects, they do resemble many fish species. However, much of the 'fishiness' of these air-breathing mammals is superficial. Fish and cetaceans are similar in body shape because of 'convergent evolution', a process whereby unrelated species come to resemble each other as a result of adaptation to their environment over many millennia. Water is a more resistant medium than air, and a smooth, streamlined body shape reduces drag forces. Cetaceans' forelimbs have become flippers, which act as hydrofoils to create lift during swimming and also to act as rudders, helping them to steer. Their large, flat tails provide the main propulsive force by moving up and down, unlike the typical side-to-side movement of a fish tail. Drag is actually greatest at or near the water surface so, to save energy, it is actually better to swim some distance below the surface. When travelling fast, some species will also use the lower resistance of the air by regularly leaping out of the water, a process known as 'porpoising'.

Cetaceans have a fat-rich blubber layer just below the skin that wraps almost their whole bodies in a thick layer of insulation. The blubber is also a vitally important energy store and a cetacean can be thought of as a 'swimming rechargeable battery'. Many animals have to feed almost constantly to survive, but after recharging their energy supplies by feeding and then storing fats in their blubber, many cetaceans can roam widely and indulge in many other behaviours, such as socializing, mating, learning and exploring, before they next need to feed. The biggest 'marine batteries' are the great whales and some of the baleen species probably do not eat for many months between their annual visits to their polar feeding grounds. Similarly, deep-diving whales use their energy-store capacity to fuel their expeditions deep underwater.

Incidentally, cetaceans – like all other mammals – also need to rest and they apparently do this by closing down one half of their brain at a time. This leaves them adequately responsive to their environment. While sleeping they continue to swim in simple patterns and also come to the surface at regular intervals to breathe.

ABOVE Here Spinner Dolphins travel at speed by 'porpoising' in the Hawaiian Islands.
LEFT The most advanced ocean predator resting – an Orca sleeping at the surface of the ocean in New Zealand.
OPPOSITE A beautiful school of Atlantic Spotted Dolphins in the Bahamas.
PAGE 64 A large, swiftly moving school of Long-beaked Common Dolphins in the Sea of Cortez, Mexico.

KEY ASPECTS OF CETACEAN PHYSIOLOGY

If you understand something of the biology of mammals (and we humans, of course, are also mammals), then you know a lot about whales and dolphins already. There are a few key points to remember about cetaceans:

- They live in water but breathe air.
- They live in a world where hearing, rather than sight, is the dominant sense, as light diminishes rapidly with water depth.
- While they have to return to the surface to breathe, their world includes the water column below, making it more three-dimensional than ours.
- They are predators – they need to find and ingest adequate numbers of other animals to survive.

Rapid loss of epidermal cells keeps most cetaceans free of the barnacles and other 'fouling' organisms that would otherwise colonize their skins. Those cetaceans that sport protuberant growths – like the huge callosities on the heads of Right Whales – must have sacrificed some of their streamlining for very good reason, given the extra energy that they need to exert to overcome the drag caused by these structures. The layer of blubber, which also provides extra buoyancy, helps give cetaceans their smooth shape.

The positioning of the cetacean blowhole (or holes) at the top of the head allows for an efficient exchange of gases (breathing) without the animal having to slow down significantly when it comes to the surface. Blowholes are the equivalent of the nostrils of land

mammals. Internally, much cetacean anatomy is typical of all mammals: they have a chest cavity, containing lungs and heart, and, separated from it by the diaphragm, an abdominal cavity where stomach structures, intestines, liver, kidneys and the other usual mammal organs are found.

Cetaceans have multi-chambered stomachs, which are similar to those of cows and other ruminant animals, and perhaps tells us something about cetacean ancestry! The elastic 'fore-stomach' receives the food from the oesophagus. It is then passed to the main digestive chamber, the glandular-stomach and thence into the 'U'-shaped stomach, which regulates the flow of digested food onward into the intestines (which are relatively long in cetaceans).

The kidneys are large and lobulate, resembling a bunch of grapes. The male and female reproductive tracts have the same components as terrestrial mammals. In females, the vagina leads to a uterus and a pair of ovaries, and in males there is a penis and pair of testes. The key difference is that the penis is retractable and normally lies within the body wall. The male and female reproductive openings are ventral and situated towards the rear of the abdomen.

BELOW Sex at sea: Mating Atlantic Spotted Dolphins. The male is the lower and less spotted of the two.
OPPOSITE A Southern Right Whale being rather closely investigated by some divers off Argentina. Note the large stony patches known as callosities on its head.

Mating is carried out 'belly to belly', raising some interesting problems for animals that must regularly come to the surface to breathe. The internal mammary glands lead to mammary slits on the females' lower abdomen near the genital and anal openings.

The testes in terrestrial mammals are typically external to the abdominal cavity, believed to be an adaptation to keep them cool. In cetaceans, they are located deep inside the abdominal cavity and are kept cool by a special counter-current blood system. The blood supply to the brain in cetaceans is also complex and this, too, may be concerned with keeping the vital tissues of the central nervous system at a steady temperature as well as maintaining the oxygen supply while diving.

The brain is relatively big, spherical and deeply folded, although the extent of this folding varies from species to species: for example, the brains of river dolphins are a little smoother than those of marine dolphins. Part of the relatively large size of cetacean brains is ascribed to the development of certain areas to facilitate their sophisticated acoustic abilities. By comparison, the cetacean brain region that deals with smell seems to be poorly developed in toothed cetaceans, although they do seem to be able to detect water-borne 'odours'.

ABOVE   Mating competition in the Mexican breeding grounds: the extended penis of a Grey Whale.
BELOW   Jojo, a wild sociable Common Bottlenose Dolphin, in an open-mouthed threat posture in the Turks and Caicos Islands, Caribbean.

ABOVE The huge flipper and head of a breaching Humpback Whale.

The skulls of cetaceans are unique because they have been 'telescoped' by evolutionary pressures to allow respiration to occur via the top of the head and to create significant forward projections of 'jaws' in both toothed and baleen whales. The jaws of toothed cetaceans produce a rather bird-like skull, with a protruding 'beak', although in some toothed cetaceans this bony beak is hidden under the soft tissues of the head. In baleen whales, the upper jawbones form a massive arched structure, allowing for the baleen filter-feeding plates to be accommodated.

The cetacean vertebral column (or backbone) is another massive structure, made up of many individual vertebrae, to which large muscle masses that allow locomotion are attached. Several of the cervical (neck) vertebrae are usually fused in cetaceans, limiting head movement, the river dolphins, Belugas and Narwhals being exceptions. Cetacean ribcages are flexible, allowing changes with pressure and lung volume. The cetacean forelimb bones are a contracted version of those seen in other mammals,

including five 'fingers' made up of many small separate bones now locked inside the flipper. Flipper mobility varies between species and the primary purpose of the flippers is steering. They are not used to help propulsion. The Humpback Whale's huge flippers are famously mobile, whereas those of the Bottlenose Dolphin are relatively rigid. The hindlimbs and pelvis are missing, but are represented by a small 'pelvic vestige' bone hidden inside the body.

The main source of propulsive force resides in the muscles along the back and tail. Large muscles attached along the top of the vertebrae contract to move the tail up, and muscles attached below the vertebrae pull it back down. The dorsal fins (where they exist) and tail flukes are unique elements of cetacean anatomy and have no bones to support them.

RESPIRATION AND LOCOMOTION

Different cetacean species spend different amounts of time away from the surface, reflecting their varying life strategies. For example, Common Dolphins forage at depths of up to 260 metres (over 850 feet) for eight minutes or more, although most of their dives are limited to around 80 metres (260 feet) and last only a few minutes. Northern Bottlenose Whales regularly dive to 800–1,500 metres (2,625–4,920 feet). They are frequently underwater for 20–40 minutes and sometimes for up to two hours. Such deep-diving animals are exploiting food that is only found at these remarkable depths.

The champion cetacean diver is probably the Cachalot, which may spend about two-thirds of its time on deep, underwater foraging missions looking for deep-sea squid. Their usual dive depths are in the region of 300–800 metres (985–2,625 feet), but they can descend to 2 kilometres (over one mile) and probably beyond on occasion. An average dive for a large adult Cachalot starts when it characteristically raises its tail flukes high in the air. The whale then makes an almost vertical descent at a speed of 30–100 metres (100–330 feet) per minute. After about 15 minutes it slows near the level of its maximum descent, forages for some 15–30 minutes and then makes another almost vertical journey back to the surface. Calves do not dive as deep and are usually left at the surface. Shallow dives not preceded by the characteristic 'fluking-up' are reported as a response to disturbance.

Whereas humans (without the aid of machines and with the exception of a few specialist athletes) can, typically, dive only to a depth of a few metres and for less than a minute, marine mammals have a range of physiological adaptations to allow for deeper and more prolonged diving. Oxygen can be stored in the lungs, blood and muscle of mammals and, in cetaceans, the blood and muscles are the primary oxygen reservoirs. This is reflected by their relatively large blood volume and small lungs. They also have high levels of myoglobin in their muscles, a pigment that allows oxygen to be stored efficiently.

Marine mammal muscles and other tissues are also resistant to exposure to high concentrations of carbon dioxide and lactic acid, which are the waste products of respiration. Cetaceans can also control their blood supply so that essential bodily functions and organs, including the brain, are still supplied with oxygen-rich blood and protected from the toxic waste products of respiration. When they dive the heart rate of marine mammals actually slows, a process called bradycardia, and other non-essential body functions do likewise, helping to slow oxygen consumption.

Another important factor is internal pressure which, within the lungs and any other airspaces, needs to be close to the external pressure, which increases rapidly and significantly with depth, to avoid damage. The anatomy of the lungs of diving mammals allows them to collapse in stages, thereby controlling the internal pressure.

Cetaceans live in an environment that is colder than their internal body temperature. Their insulating blubber layers

ABOVE Diving in cetaceans is primarily concerned with finding food. The deepest divers have evolved to exploit resources far below the surface, such as the deep-sea squid that form the main part of the diet of the Cachalot or Sperm Whale. Here a Cachalot raises its massive tail high in the air as it leaves the surface for a deep foraging dive in the Gulf of California, Mexico.

OPPOSITE Northern Bottlenose Whales – an adult and calf (note the darker colour and shorter beak) – seen in one of their most important habitats, The Gully, off the eastern coast of Nova Scotia, Canada.

help to maintain body temperature, as do the 'counter-current exchangers' in their circulatory systems. For example, the surrounding water cools the blood near the surface of the dorsal fin. This blood then passes through a system of vessels, and blood closer to core body temperature runs across the vessels containing the cooler blood. In this way, heat is exchanged between the two. A similar system also allows cetaceans to lose heat through their extremities, helping to keep body temperature within the normal range. Furthermore, cetaceans can also lose blubber and become thinner in the warmer months and fatter in the winter, when they need more insulation.

ABOVE   Fast moving Orcas surfacing off Patagonia. The Orca typically co-ordinates its hunting activities and lives in close-knit family groups.
RIGHT   Atlantic Spotted Dolphins in the Bahamas playing with a bubble-ring that one has produced. In addition to manipulating solid objects, some dolphins are also quite creative with air bubbles.

FEEDING – THE GRAZERS AND THE HUNTERS

The toothed cetaceans hunt and ingest large – usually single – prey. Their teeth are not used for chewing, but to catch and hold prey. By contrast, the baleen whales filter feed, typically catching large numbers of small animals in each mouthful. Their baleen plates, sheets of keratinized epidermis, similar to hair, fingernails and horn, are suspended from the roof of their arched mouths. In some species, the bones of the lower jaw are loosely connected and can be spread far apart to allow large mouthfuls of water and prey to be engulfed, a process assisted by elastic throat grooves. The different feeding techniques of the baleen whales are described on page 41.

Most cetaceans roam great distances in search of food. For some species, this means making remarkable migrations. For others, it requires shorter lateral movements from one patch of prey to another, or significant regular, vertical movements to exploit deep-water species. Presumably those animals making long

BELOW The partly opened mouth of this Humpback Whale in Alaska reveals the comb-like baleen plates hanging from the upper jaw. A swimming rorqual whale, such as this one, will open its mouth as it approaches a dense patch of prey, causing water to rush in, helped by the forward motion of the swiftly moving whale and the huge expansion of the throat. After the mouth is closed, water is forced through the filter plates of baleen by the tongue and the contraction of muscles in the throat grooves, and the prey is captured on the baleen plates.

migrations to polar feeding grounds are following instinct that is sparked by seasonal changes.

'Echolocation' (see page 82) clearly plays an important part in the detection of prey by many cetaceans. The sea is actually a rather noisy place and the acute hearing of cetaceans may mean that they can sometimes hear and locate their prey without even producing echolocation sounds. Indeed the Orcas that specialize in hunting other marine mammals choose not to produce echolocation sounds when hunting so as not to warn prey of their presence.

Cetaceans are known to use various other hunting techniques, including stealth and herding, reflecting their local conditions and the nature of the prey. Herding fish, plankton or squid can be achieved by the co-ordinated movements of members of cetacean groups approaching

ABOVE A sociable Common Bottlenose Dolphin in the Red Sea plays with an octopus thrown to it by a fisherman before eating it.
LEFT Long-beaked Common Dolphins feeding on a shoal of sardines off South Africa. Cetaceans often herd their prey, causing fish (and also sometimes plankton and perhaps even squid) to form concentrated prey balls, which they can then feed from efficiently.

prey from different angles simultaneously, or by pushing them towards natural boundary features such as the shore or even the surface of the water. Dolphins are also known to strike fish with their tails to incapacitate them, and there is mounting evidence that they may also be able to use their remarkable acoustic abilities to stun prey by 'firing' pulses of sound at them.

Typical toothed-cetacean prey, such as fish and squid, is fast-moving and highly reactive. This is also true when the prey species is another marine mammal. Mammal-hunting Orcas kill and eat prey that are both smaller than themselves, for example seals, and those that are far larger, such as Grey Whales. In the case of the larger animals, the Orcas rip away chunks of flesh and may eat only part of the carcass, notably the blubber and the tongue, which are particularly energy-rich. Like certain

terrestrial predators, Orcas may direct their attention at the most vulnerable group members, such as the young. Some large whales have defensive responses when faced with a hunting pack of Orcas. Adult Cachalots form a defensive ring around the vulnerable young members of their group, with either their large heads or powerful tails pointing outwards. They may also take it in turns to split away from this defensive formation to respond to the attacks. Orcas have other ingenious hunting behaviours, such as knocking their prey off floating ice or rocks by moving their bodies violently to create powerful waves nearby. In one location, on the Patagonian coast of Argentina, they beach themselves temporarily to reach the Southern Elephant Seals and Southern Sea Lions that haul out to breed on the steeply sloping shores. The whales have learnt to stalk their prey from the shallow water and then, at the crucial moment, hurl themselves up onto the shore to grab their hapless victim. They then twist around and shuffle their bodies

back into the water. High-risk activities like temporary beaching are clearly a skill learned over time with little room for error. Older Orcas pass on the technique to younger animals and the Orca calves in this Patagonian population can be seen practising this activity before they try it for real. Despite their fearsome reputation, there is no hard evidence that Orcas have ever hunted or killed human beings.

Another remarkable feeding pattern is the bubble-netting technique used by the highly adaptable Humpback Whales. Fish and krill avoid bubbles and when Humpback Whales encounter a dense 'swarm' of prey, one or two individuals manoeuvre below them and then swim upwards in a spiral motion emitting a train of bubbles from their blowholes. These bubbles cause the fish and krill to clump together and the whales then swim through this mass with their mouths open. Observers may witness a ring of bubbles erupting at the surface as evidence that Humpback Whales are indeed bubble-

netting. Nor are these the only baleen whales that collaborate when feeding. Bowhead Whales are reported to line up in a 'V' formation, which may help to funnel more prey to each individual.

In addition to food, cetaceans need water and appear to obtain this from two sources: their prey, for example fish, which are a handy 'pre–packed' supply, as fish tissues are some 60 to 80 per cent water, and also the process of the digestion of fats, carbohydrates and proteins, which releases water. The same is true of the metabolism of stores of fats and proteins. Marine mammals also show some physiological adaptations – similar to those seen in desert animals – that probably help them to conserve water, such as the production of very concentrated urine.

ABOVE Common Bottlenose Dolphins in South Carolina, USA, have learnt to drive fish onto the shore and then, quite remarkably, partially strand themselves to catch the concentrated and incapacitated fish.

LEFT This group of feeding Humpback Whales in Alaskan waters has used bubbles to concentrate their prey in a technique known as 'bubble-netting', which is described on page 78.

OPPOSITE The open mouth and baleen plates of a Grey Whale feeding amongst the kelp off California.

REPRODUCTION AT SEA

Cetaceans have a relatively long lifespan and produce low numbers of young. Their pregnancies are long and, in some species, suckling continues for several years. Extended periods of maternal care are typically followed by slow maturation.

All species of marine mammals normally produce single offspring, probably because their young need significant and dedicated care in order to improve their chances of survival and mothers can probably only produce enough milk to feed one youngster at a time. Nonetheless, cetacean calves are born well developed and can swim immediately. Suckling under water is achieved when the calf fixes on to one of the mother's two teats, which sit in slits on her belly, and milk is then squirted into the youngster by special abdominal muscles. The milk of cetaceans is exceptionally rich in fat and protein and calves grow rapidly.

The variations in the lactation period between species reflect differing ecological conditions. However, the presence of a suckling youngster does not necessarily mean that the mother has been feeding it continuously throughout its life. Juveniles of some species that are already feeding themselves effectively on fish may still occasionally suckle and not necessarily just from their own mother.

The nature of the prey that each cetacean species depends primarily upon strongly influences their reproductive cycles. Cephalopods (squid and octopuses) have a lower nutritional value than most fish and cetaceans that prey on them tend to have longer pregnancies and suckling periods. Very few cetaceans give birth annually and most produce a calf every two to three years. Pilot Whales, Cachalots and Orcas breed even less frequently. Low breeding rates and late maturation mean that females only produce some 5–25 young during their lifetimes. Older cetaceans typically produce fewer young but, in some species at least, these 'senior citizens' play a role in assisting in the care of the young of others. Some Short-finned Pilot Whale females have a post-reproductive life that can continue for several decades and may even foster the young of other females.

ABOVE Pink and grey: the contrasting colours of an Indo-Pacific Humpbacked Dolphin mother and her calf in Hong Kong.

LEFT Big baby – A Southern Right Whale mother and her calf in Argentina.

OPPOSITE Dolphin calves usually spend their time in the 'follow position' swimming just below their mothers. This youngster – a Pantropical Spotted Dolphin – appears to prefer, instead, to ride on its mother's back – which is very unusual behaviour.

ECHOLOCATION

Even close to the surface of the sea, cetaceans – many of which have good eyesight – may be unable to use their vision because of the murkiness of the water. Similarly, light rapidly declines as they dive and vision again becomes compromised. Many cetaceans have, therefore, evolved a very acute sense of hearing and a sensory mechanism known as 'echolocation'. Basically, echolocating cetaceans produce high-frequency 'clicks', which bounce off the seabed, underwater structures and other animals around them. Their brains have developed to provide a detailed analysis of the echoes of their clicks. A similar process is used in the sonar systems that humans deploy, one type of which is now used by some fishermen to identify schools of fish of different species. Cetacean echolocation is far more sophisticated even than this, and allows them to navigate accurately and to find food even where there is no perceivable light. Experiments have shown that the Bottlenose Dolphin can use its remarkable acoustic abilities to detect small targets accurately even up to 100 metres (over 300 feet) away.

The echolocation click is emitted as a beam of sound from the front of the cetacean's head. How echolocation and other sounds are produced varies from one species to another to some extent and is best known for the Bottlenose Dolphin. The head of a dolphin contains a number of complex structures not seen in terrestrial mammals:

- Between the blowhole (at the top of the head) and the bones that form the skull, is the bulging 'forehead' area that contains the structure known as the 'melon'. This is a lipid-rich tissue that acts as an acoustic lens, helping to focus the sounds emitted by the animal.
- Behind the melon is a series of structures branching off from the 'nares' (the two tubes connecting the blowhole to the lungs) known as nasal plugs. These structures include various muscular plugs and air tubes: amongst these are
- The 'Phonic lips', which many scientists believe are used to produce both echolocation clicks and other noises. Other scientists still favour the larynx, which is used to produce sounds by many terrestrial mammals, as the source of cetacean sounds.

OPPOSITE   An adult Fin Whale surfacing in the Sea of Cortez. This head-on view shows the 'splash-guard' that protects the blowholes, which are just behind it.

ABOVE   Two Melon-headed Whales breaching. Although their skulls are like those of dolphins, an external beak is absent in this species because of the extensive fleshy 'melon' tissue of the forehead region.

The means by which cetaceans receive sounds are similarly unique. Bats – the other order of mammals that have the ability to echolocate – have large external ears. Most land mammals and a few marine species as well have obvious external earflaps that help to focus sound into the middle and inner ears where it is processed. Not only do cetaceans have no protruding earflaps, they also have no open passage between the outside of the head and the ear structures inside. In the toothed whales, the earhole – or 'external auditory meatus' – is replaced by fibrous tissue and in baleen whales, a waxy plug, which can show growth rings, fills this same space.

The latest theory is that cetaceans hear through their chins! The bone in the lower jaw is thin and contains a discreet body of fat that conducts sound to the inner ear.

The physics of sound in water are significantly different to those in air and, unlikely though the jaw-hearing route might appear, it is supported by considerable research.

The Cachalot probably shows the highest degree of development of echolocation anatomy. An enormous acoustic apparatus, the 'spermaceti organ', takes up between a quarter and a third of the animal's entire body. Essentially, this is a huge area of soft spongy tissue, which sits between the front of the skull, which is shaped like a huge bowl and may act as an acoustic reflector, and the

BELOW Foraging Atlantic Spotted Dolphins using echolocation to help find fish buried in the sand. Their echolocation skills mean that they can find prey that they cannot see.

front of the head. A large air sac separates the skull from the spermaceti organ, with another right at the front of the head. Between the spermaceti organ and the upper jaw sits the 'junk', another huge tissue saturated with oil. The blowhole is towards the front of the head and the air passages leading to it take two separate routes. One, the right nasal passage, runs between the junk and the spermaceti organ; the other passes through the spermaceti organ.

It has been suggested that sound is initially produced by air being forced into a valve-like structure, associated with the air sac at the front of the head, known as the museau du singe (the monkey's lips). This sound pulse passes backwards through the spermaceti organ. It is then reflected back off the air sac at the front of the skull and travels through the junk and the spermaceti organ and out into the sea. The passage of sound through these two different tissues is such that each original pulse produced by the Cachalot is emitted as two or more clicks: the first one coming from the junk and the second from the organ above it slightly later. The Cachalot probably produces the most powerful of all cetacean clicks, and it appears that this species has evolved a structure that allows it to 'see' extremely well and over significant distances, even in the darkest depths of the sea.

BELOW The huge square-ended front of the head of a Cachalot, seen here in the Gulf of California, contains the soft and oil-rich tissues known as the 'junk' and 'spermaceti organ'. These are the acoustic lenses, which the whale uses to focus its powerful echolocation clicks.

LANGUAGE AND COMMUNICATION

Many birds and mammals issue warning cries and have some way of advertising themselves acoustically as suitable mates. However, for some mammals, more complicated communications also occur. Part of our problem in this area lies with our limited ability to perceive and interpret communications that are unlike our own. For example, we have almost always focused on those sound frequencies that we are able to hear, while communication may be ongoing at frequencies beyond our own hearing range.

Cetaceans are known to produce a range of calls. Some have been identified as greetings, others as alarm calls. Dolphins produce whistles that are unique to individuals and may act to identify them. These whistles may also indicate who is related to whom, like a family name, and introducing yourself might be seen as the most basic component of dolphin language. Several researchers

ABOVE   Dolphins may communicate in many different ways. Some communication may be tactile: these Spinner Dolphins are gently touching pectoral fins.

have tried to encourage dolphins to talk to them and have made efforts to persuade the dolphins to mimic human speech. Bottlenose Dolphins did not do very well in these experiments, but fared better when sign language was used, along with computer-generated sounds. They understood simple sentences and novel rearrangements of words, including some appreciation of sentence structure.

However, it is possible that these studies missed the other means of communication that cetaceans may use more frequently. People have reported feeling the echolocation clicks of dolphins in the water and at short distances dolphins may be able to 'touch' each other with

this ability and pass on information. Wild dolphins also show a wide range of signalling behaviours in their natural habitat. For example, some Risso's Dolphins off the west coast of Scotland reacted to a close-passing survey vessel with a rich display of responses. They crashed their tails against the surface (belly up and belly down), they half and fully breached and also slapped the water with

ABOVE   A whistling Atlantic Spotted Dolphin. This is another of the highly social dolphin species that is usually seen in groups.

their heads. Other members of their group could have seen and heard this behaviour over quite some distance. Perhaps they were calling the group together, or suggesting that the scientists should move on. Sounds and actions may also have different meanings according to the context in which they are made. Until we can travel with these animals underwater, monitoring closely what they are doing, including the contexts of their behaviour and, simultaneously, monitoring all their possible communication channels, it will remain very difficult to interpret cetacean communications.

Nonetheless, years of painstaking field research are yielding important interpretations of the noises of whales.

LEFT The massive 'opera star' of the oceans – a singing Humpback Whale in the Pacific near Mexico.
BELOW A social group of Cachalots off the Azores – when not on deep foraging dives, female Cachalots and calves may often be found gathered in groups at the surface.
OPPOSITE A school of socializing Spinner Dolphins off Big Island, Hawaii.

For example, every adult male Humpback Whale sings a particular song that is unique to his population. No two populations have the same 'theme tune', and early in the breeding season all the males sing pretty much the same song. As the breeding season progresses, the song subtly changes. At the end of the breeding season, singing stops until it is breeding time again when the males resume singing the same version of the tune where they left off.

The noises produced by Cachalots are particularly interesting because these animals seem to specialize in clicks, which are used for both echolocation and communication. The interval in a pulse of clicks is also proportional to the size of the whale itself. A pulse of several clicks followed by a pause is known as a 'coda' and, whilst genetic studies have shown little variation between populations of Cachalots in different oceans, the groups of females found in different regions have distinctly different coda repertoires. This has produced the notion that there are clans of Cachalots: social units with very similar coda repertoires or, to put it another way, common dialects.

In addition to the codas, the whales emit slow clicks and 'creaks'. In these creak noises, the clicks are tightly packed together and creaks seem to be used when the whale is investigating something in detail. Sometimes, one Cachalot will emit a coda and another will appear to respond with its own coda. The codas can vary in the arrangement of clicks and short pauses. There is clearly potential for a language in this, but it should be noted that solitary whales also emit codas – perhaps they are talking to themselves.

The significance of the limitations of our knowledge of the sensory abilities of other species has become apparent in the case of African Elephants. Well-established as social animals, on the basis of long-lasting herd structures, their vocabulary, nonetheless, appeared to be quite limited and one mystery was how they managed to co-ordinate the herds such that, in certain seasons, small family groups would all join up together. The answer seems to be that they can 'hear' low-frequency sounds through their feet, including the movements of other herds. Their feet have vibration sensors, known as 'Pacinian corpuscles', and an elephant standing still for no apparent reason may well be listening through the ground. There are also vibration sensors in their trunks and researchers are revisiting their low-frequency 'tummy' rumbles, which may also be communication rather than indigestion. The elephant example illustrates that, in the case of sophisticated mammals, we need to consider the possibility of communication using senses beyond the ranges of our own.

Consideration of the range of methods that humans use to communicate – words, gestures, postures, expression and direct contact (noting that these forms of communication can have different meanings in different contexts) – provides an insight into how other sophisticated mammals might also behave.

CETACEAN SOCIETIES – THE SOCIAL STRUCTURES

Cetaceans are typically 'social animals', as they live in groups of co-operating individuals, and much of our current knowledge is based on the surface behaviour of cetaceans in a school, which we can often see are clearly interacting and collaborating. However, cetaceans may be interacting over much greater distances; for example, Blue Whales may be in acoustic contact across entire ocean basins. Cetaceans may also be socializing underwater where we cannot observe them.

Group living provides an advantage in dealing with predators, finding food and rearing and protecting young. Cetacean group size may vary over time, but it is clear that some species tend to form groups of a certain size. For example, beaked whales and river dolphins are generally found in small groups of just a few individuals; whereas Pilot Whales are commonly seen in groups of hundreds or even thousands of individuals, although

around 50 may be more normal. Group size may also reflect feeding specialities, with the larger groups feeding upon abundant prey such as salmon and smaller groups taking less plentiful fish.

We know that the most basic relationship within cetacean societies is the strong bond that exists between mother and calf, but other affiliations have also been revealed in the more studied species. Pilot Whales seem to have a very stable school structure, with each school typically including adult males and females and younger animals and calves. Orcas also have a remarkably strong social cohesion within their pods, which are stable through successive generations. A typical Orca pod appears to contain individuals of both sexes, including several adult females and males.

The social structures of other marine dolphins, in so far as they are known, seem more fluid. Animals may be seen within groups, but the composition of these groups

ABOVE   Both species of Pilot Whale are highly social and often seen in large groups. The warmer water species is the Short-finned Pilot Whale, seen here off Hawaii.

RIGHT   Three kings: the towering dorsal fins of three large adult male Orcas in British Columbia easily distinguish them from the rest of their pod.

OPPOSITE   The scarred skin of a Risso's Dolphin clearly marks it out from the Bottlenose Dolphins that it is travelling with. These species seem to associate quite commonly with each other.

can change rapidly and there can be many links of differing sorts between individuals. One characteristic of the Bottlenose Dolphin society, for example, are male allegiances that form to guard mature female dolphins from the advances of other males.

Cetaceans are sometimes also seen in schools of mixed species. For example, Pilot Whales and Bottlenose Dolphins often swim together or with other cetacean species. There may be strength and security in numbers out in the open oceans.

Some cetaceans can even form mutually beneficial partnerships with non-cetacean species, including our own. For example, for over 100 years, and three human generations, one particular group of Bottlenose Dolphins in Laguna, Brazil, has been driving mullet towards fishermen. This activity is now so well co-ordinated that, with a certain splash of their tails, they direct the fishermen when to use their throw-nets. The dolphins then feed on the remaining fish that have been concentrated between them and the men. As long as the weather is suitable, this occurs on a daily basis and dolphins sometimes even go to look for the fishermen to encourage them to go fishing. Similar human–dolphin-fishing teams also occur in other parts of the world, where they have separately evolved.

WHALE CULTURE

It is only very recently, as field studies have developed, that it has been possible to recognize that cetaceans have such sophisticated behaviour that it can be termed 'culture', meaning a knowledge passed from one generation to the next. An example would be the Patagonian Orca that feeds on sea lions by partially stranding themselves. Another example would be the use of natural sea sponges by female Bottlenose Dolphins in certain pods off Australia, and possibly elsewhere, apparently to protect their sensitive beaks when foraging on the seabed. Culture can no longer be considered to belong exclusively to human societies. Indeed, scientists are now suggesting that the complex and stable cultures of the Orcas, at least, have no parallel outside human societies.

LEFT An Orca dramatically comes up onto the shore to try to snatch a South American Sea Lion. This highly specialized hunting technique is peculiar to one population and passed from one generation to the next.

Life with a big brain

HOW INTELLIGENT ARE CETACEANS?

Whales and dolphins are animal record breakers in many anatomical and other physiological respects and we are particularly fascinated by their big brain sizes. We know – and take pride in the fact – that humans have far larger brains than most animals, and when we ponder the intelligence of other species, we typically ask about their brain size first. The brain of the Bottlenose Dolphin is a little larger than our own. The Orca's brain is about four times the mass of ours and the Cachalot actually owns the largest brain on planet Earth – six times the size of the human brain, more than 9 kilograms (almost 20 pounds) of grey matter.

However, brain size alone does not relate directly to brainpower. Brain size relative to body size was first suggested as a better guide in the 19th century, which helped to put the Cachalot in its place. More recently, the area of the brain surface and the development of certain regions of the brain, such as the neocortex, were put forward as better indicators of brainpower. Differences in brain anatomy between cetacean and terrestrial mammals reflect some 50 million years of separate evolution, and a significant part of the cetacean brain – particularly their highly specialized auditory system – is clearly concerned with aquatic adaptations. These brain structures also suggest a form of intelligence different to our own. However, the study of brain anatomy alone does not give satisfactory answers to the question of intelligence.

We value intelligence in our own species, recognizing it as the main attribute that allows us to survive and be successful. We also tend to value animals for their intelligence and feel far more concerned about harming those we deem to be intelligent. Some of this concern is based on the premise that we share similar feelings of pain, fear and loss, combined with a growing appreciation that we may also share a context of families and societies. We appreciate that killing an intelligent animal may have important immediate implications for its family, social group or even its population. The animal that we remove from the wild may be a mother, a 'teacher', or some other part of an important social unit.

It may be carrying some knowledge about resources or threats that, at some point in the future, it would have usefully shared with other members of its population. It is not surprising, therefore, that those involved in the whaling issue often debate cetacean intelligence.

Our understanding of dolphin and whale behaviour is estimated to be more than 30 years behind our appreciation of that of non-human primates. Yet, we can already appreciate the keen curiosity and playfulness of the Bottlenose Dolphin and tests have confirmed its self-awareness. In addition, field studies reveal excellent examples of advanced group co-operation, for example, when rounding up fish or driving off predators, as well as complex relationships within populations. In the open ocean, when challenged by predators or otherwise frightened, cetaceans have nothing to hide behind but each other: a factor that has clearly profoundly shaped the evolution of their co-operative behaviour and intelligence.

Intelligence is not easy to define. It may relate to the ability to store facts, use reason or respond rapidly in an appropriate way. It is even more difficult to judge in animals that live in a very different way to us and with which we have few common points of communication. In fact, the history of the study of intelligence in cetaceans is also inextricably linked to investigations into their communication. Nevertheless, it is clear that some cetaceans demonstrate:

- The ability to use tools.
- Problem solving and abstract concept formation.
- Self-awareness and highly responsive behaviour.
- Shared communications and social behaviour.
- The ability to work co-operatively to achieve their goals.
- A high degree of vocal adaptability.

ABOVE A Humpback Whale slaps its long pectoral fin against the surface in Monterey, California.
RIGHT A Dusky Dolphin displays at the surface of a glassy sea off Kaikoura, New Zealand.

The local gang and the great migrators

Cetacean species inhabit all parts of the seas and oceans and can be found thousands of kilometres from land, in the coastal zones, in lagoons, bays and estuaries and a few species have evolved to live in some large river systems found in the tropics. Some, for example, the Narwhal and the Bowhead Whale, are only found in cold waters. Others, such as the Tucuxi, exclusively inhabit tropical waters. Orcas are 'cosmopolitan' and are found in all marine areas but are also a 'species-complex' (comprising more than one species).

The strong relationship between cetacean distributions and water temperature zones may be because cetacean species seek particular temperatures or because their favoured prey are found there, or both. The map on pages 148-9 shows the major climate zones of the seas, illustrating where tropical, temperate or polar species may be found. Overlaid on these temperature zones are the other habitat preferences of the cetaceans such as depth and salinity, some being only found in the coastal zones, others further offshore, and also the migration routes that take some whales species through several climate zones.

BELOW The extraordinary 'tusk' of a surfacing Narwhal.

Not surprisingly, many cetaceans tend to be found in areas of high productivity, such as those where nutrients are brought up from the deep seas by upwelling currents. The seas are far from being homogenous water bodies: they contain various mobile frontal zones, where water bodies with differing physical and chemical natures meet. These also tend to be areas where both marine predators and their prey are often particularly plentiful.

Some cetaceans clearly have a limited range, whilst others roam over greater areas and appear, at least at first, to be more adaptable. However, an apparently wide range may hide the fact that species may only be exploiting certain aspects – such as particular depths of water or marine features – where a particular prey may be found. Risso's Dolphin, for example, is found in many areas, but a careful examination of its distribution shows that it is associated primarily with deeper areas where this deep-diving species finds its prey.

The baleen whales are usually thought of as being highly migratory, with exceptions such as the Arctic-dwelling Bowhead Whale. Certainly, the Grey Whale is amongst the most migratory of any animals and famously follows a shallow coastal route in the North Pacific. At one end of its migration it breeds in coastal lagoon systems, whose warm waters form nursery areas for the calves, and at the other end is its main northern feeding ground. Bryde's Whales are a baleen species that do not appear to make long migrations but live year-round in the warm waters near the Equator. However, some or all may make seasonal movements, following their prey within the latitudes where they live. The Common Minke Whale in the North Atlantic may do something similar and perhaps such populations should be described as 'weakly migratory'. It has recently also become clear that the famous long migrations of some of the other large baleen whales – the Blue, Fin, Sei and Humpback – are less predictable than once thought. For example, the use of ex-military listening stations in the northeast Atlantic has revealed that some of these whales are present off the British Isles year-round. In the summer of 2003, whale-watchers off the Azores happily encountered large and friendly groups of Fin Whales, when they might have been expected to be far away to the north. At much the same time, another whale-watching expedition off Iceland was bemoaning the unusual lack of large whale sightings there. The take-home message may be that whereas we have a pretty good idea of what some of the whales do most of the time, we only have a poor idea of what most of the whales are doing at any particular time.

OPPOSITE   A surfacing Grey Whale off Mexico.
BELOW   A large school of passing Belugas in the Canadian Arctic.

Chapter Three

MAN AND WHALE

What is a whale worth?

Our relationships with cetaceans – both historical and modern – range from admiration and inspiration to exploitation and callous disregard.

Over the centuries, the human species has benefited materially from the existence of the great whales. Firstly, they were a primary food resource. Then they became the source of some of the products that helped to drive our industrial revolution. Most recently, they have become the focus of the growing and lucrative whale-watching industry that sits at the heart of marine eco-tourism. It can also easily be argued that we have benefited spiritually from our interactions with the inspiring giants of the sea, as reflected in art and literature.

In this chapter, we start to delve below the surface of our historical and modern interactions with these animals to explore a little of what lies beneath.

A short history

Mankind and whale-kind have a long joint history, although the cetaceans have been on Earth, pretty much in their present modern form, much longer than us. Fossils indicate that the modern cetacean species were present from around the Middle Pliocene to Early Pleistocene periods, 3.5–1.8 million years ago, whereas *Homo sapiens* originated only some 300,000 years ago.

Some of the cetacean–human interactions seen today could originate from the time when humans first established coastal communities. The curiously friendly relationship between dolphins and humans is certainly not new and is celebrated in early myths, legends and art. Some legends suggest that the benevolent nature of dolphins towards humans results from the fact that they were themselves human beings, before being transmogrified into their current dolphin forms. Peoples living along the tributaries of the Amazon River in South America tell a rather different story. In a local legend, the Boto or Pink River Dolphin emerges from the river in the

form of a red man and his visits are conveniently used to explain unexpected pregnancies in the local population.

Dolphins are also featured on many early Greek artefacts and in many stories. The most famous legend concerns the poet and inventor, Arion, said to have lived in Corinth in the late 7th century BC. Having been thrown into the sea by pirates, he then charmed a dolphin with his music and the dolphin carried him, on its back, safely to shore.

Arion's story may be the first reported example of dolphin rescue, but it is certainly not the last. Accounts of dolphins aiding humans in distress come in two forms: firstly as reports of drowning people being helped to the surface or even the shore by one or more dolphins and, secondly, as incidences where dolphins have driven off sharks. The question has often been asked whether the dolphins set out deliberately to help, or if their actions are purely instinctive. Certainly, dolphins protect their own schools by attacking sharks and driving them

away. Humans in the water nearby could be forgiven for thinking that such 'brave' actions were being undertaken for their benefit, whereas the dolphins might simply be protecting their own kin. Alternatively, it could be that the dolphins recognized the vulnerability of the swimmers and responded, or that their concept of the need to protect their school expanded to include the humans.

Dolphins also have a strong instinct to protect weaker members of their schools. There are many reports of sick, wounded and dying animals being helped to the water surface by others to enable them to breathe. Could this help to explain the many accounts of drowning people claiming to have been saved by dolphins? Do the dolphins recognize the circumstances and deliberately help, or is this merely an instinctive response? We shall probably never know for certain.

Nonetheless, several people around the world are confident that they have been saved from drowning

ABOVE   Jojo, a sociable dolphin – meaning that he often chooses to associate with human beings – plays with a Nurse Shark. Sociable Bottlenose Dolphins have also been termed 'ambassador dolphins', suggesting that they are acting as ambassadors for their own kind or perhaps marine wildlife more generally. Sociable dolphins, like Jojo, in the Turks and Caicos Islands, Georges, in France and southern England and, Fungi, in Ireland, often become famous and our enthusiastic interactions with them need careful management.

OPPOSITE   A leaping Boto or Pink River Dolphin, whose pink skin colour helped to spawn the legend of the dolphin that becomes a man by night and leaves the river.

PAGE 100   Once the focus of industrial whaling activities, here a Cachalot is inspected by a friendlier human visitor off the Caribbean island of Dominica, .

or shark attack by the intervention of dolphins. One extraordinary account comes from a group of fishermen from South Carolina, USA. In June 2001 they were fishing some 56 kilometres (35 miles) off the coast of Georgetown when their boat sank. The men drifted with the Gulf Stream and later described their situation in a letter to a local paper as 'surrounded' by Mako, Hammerhead, Tiger and other sharks. The attention of the sharks was so great that the fishermen were scraped and bruised by them. Then, fortunately, a group of dolphins arrived and drove the sharks away. Even more remarkably,

ABOVE   Here an adult Common Bottlenose Dolphin is seemingly 'adopting' a Spinner Dolphin calf. This image was taken off the Tuamoto Archipelago in French Polynesia.
OPPOSITE   A small group of Clymene Dolphins flying through the air as they travel at high speed in the Gulf of Mexico.

the dolphins stayed with the men throughout the night and, the following afternoon, were still there to repel what the men reported as an attacking 2.7-metre (9-foot) Great White Shark.

Evolving attitudes to marine mammals

A few years ago, the Whale and Dolphin Conservation Society (WDCS) sponsored a professional polling organization to carry out an opinion poll of British attitudes to cetaceans. The results revealed a remarkably strong response to whaling. The vast majority of the British public objected to whaling for meat, blubber or other purposes, with seven out of ten stating that they strongly opposed it. In stark contrast, Icelandic sources have suggested that whaling is supported there by 70–80 per cent of the population.

As these examples show, there are clearly radically different opinions between peoples and countries about marine mammals. At the root of this may be fundamental differences in the way that people view the animals concerned. For example, if you truly believe that expanding whale populations are threatening your strongly fisheries-based economy, you may view them less benevolently than those who consider them delightful and inspiring, or an integral and desirable part of marine ecosystems.

In many countries there has been a movement away from regarding cetaceans from the purely 'utilitarian' perspective as food, competition or a nuisance. This shift reflects a new vision that their social nature, intelligence and self-awareness should bestow some intrinsic rights. We also now increasingly recognize that these animals can suffer in ways which we can comprehend and with which we can empathize. For example, the welfare concerns associated with whaling have now been on the agenda of the IWC for a number of years, despite apparent reticence from at least some of the pro-whaling lobby.

However, the development of a more benevolent approach to marine mammals amongst the general public in many western countries has also spawned something of a backlash. A strong, well-funded 'pro-utilization' lobby has developed in response and a growing number of organizations work on behalf of these interests, including pro-whaling, non-governmental organizations.

Our interaction with Orcas over the years provides an illustration of our changing relationships with cetaceans. A few decades ago, their fearsome reputation as 'killers' and plunderers of fish stocks prompted military expeditions to exterminate them. Today, however, killing Orcas in the name of fisheries protection has become very rare and a new type of commercial interest has taken over. Now, hundreds of thousands of people every year join commercial whale-watching expeditions to view and photograph these animals in the wild. Similarly, tremendous efforts were made on behalf of the welfare of even a single individual: Keiko – the Orca who starred in the Free Willy films and who died in 2003 – for example, was the subject of a pioneering programme to reintroduce him to the ocean. Our interest and enthusiasm for Orcas and other cetaceans is so great that many people will now pay large sums of money to see, pet and swim with these animals in zoos and aquaria. This provides an incentive to hold them in captivity and maintain the trade in animals caught in the wild.

In February 2000, a wild Orca seemed to stage its own 'anti-captivity rally' deep in the heart of Japan, where there are many dolphinaria. The lone male unexpectedly swam into the harbour of the city of Nagoya, which is also known as the 'City of the Golden Orca'. He rapidly

ABOVE   A snorkeller joins some Atlantic Spotted Dolphins on the Little Bahama Banks.
LEFT   Who is watching who? A spy-hopping Humpback Whale considers some visitors in Hawaii.

became a major national celebrity, travelling along a canal that runs through the heart of the city watched by large crowds of people, before he was encouraged back to sea some three days later. His visit to Nagoya was all the more fascinating because it coincided with much controversy over plans to build new tanks to hold Orcas at the city's famous aquarium.

There is insufficient scope here to detail the full arguments against keeping cetaceans in captivity, or to weigh them against the counter-arguments. WDCS believes cetacean captivity and the removal of animals from the wild to be cruel and unnecessary and to pose a conservation threat. For example, the 2002 meeting of

the Conference of Parties to CITES (Convention on the International Trade in Endangered Species) considered this issue in the case of the diminished Black Sea population of Bottlenose Dolphins. While captures for dolphinaria from this population were not regarded as the biggest threat, it was agreed that they should be addressed and strict new controls on international trade were agreed.

Although live captures for aquaria are banned in some countries, in others, such as Cuba, the Solomon Islands, Japan and Russia, they still continue. Dolphins and other marine mammals can even be ordered from websites, where the names of species are matched against price lists.

Man and whale   107

LEFT A school of Atlantic Spotted Dolphins in the clear open waters of the Bahamas – free to travel where they will.
OPPOSITE ABOVE Long-beaked Common Dolphins, flanked by Cape Gannets, race towards a feast of sardines off Transkei, South Africa.
OPPOSITE BELOW Jojo, a wild sociable Common Bottlenose Dolphin in the Caribbean, is fascinated by propellers, despite many resulting injuries.

While the regulation of dolphinaria and cetacean captures has become tighter in some parts of the world, such as the USA and Europe, captive facilities that are not subject to such restrictive management requirements have mushroomed elsewhere. This is the case in the Caribbean, where there is also a rich flow of tourists willing to pay large sums of money to see and swim with dolphins.

It may be salutary for anyone visiting a dolphinarium, no matter how well it is run and how good the care that is lavished on its inmates, to reflect on whether they are indirectly helping to maintain a worldwide demand for wild-caught dolphins and other cetaceans. Not only will such animals be removed from their families and social groups, but also other cetaceans may be harmed and killed during the captures. Live captures typically involve corralling schools and netting them. In Japan, live dolphins bound for captivity might be considered the 'lucky ones' because they are taken from the same drive hunts that are used to catch dolphins for meat. It is likely that the money made from sales to the captivity industry is helping to keep these drive fisheries going.

The dolphin as doctor

Encountering wild whales and dolphins can certainly be an exhilarating experience and, over the last few decades, a worldwide interest has developed in the therapeutic powers of dolphin encounters. Studies, typically using captive dolphins, provide some evidence that autistic children and people with other similar disorders may benefit from exposure to dolphins. Although there is no evidence to suggest that other forms of animal therapy – such as spending time with pets and other domesticated animals – are any less effective. This has blossomed as a treatment known as Dolphin Assisted Therapy (DAT). People also seek out wild whales and dolphins for 'spiritual enrichment'. This is usually linked to a personal belief in their mystical powers. Arguably this is not a new development; the animal spirit guides of some peoples, such as some native American tribes, have long included cetaceans.

Interest in contact with live cetaceans has also meant a global expansion in the scale of the 'swim-with' programmes, where paying customers get into the water with the animals. These programmes are now commercially offered both with wild cetaceans out at sea and with fully captive animals. A third category uses dolphins that are able to swim in and out of the encounter areas into the open sea. These dolphins are usually fed in the encounter bays; unfortunately there is increasing evidence that 'provisioning' (hand feeding) of dolphins makes them less able to survive in the wild. For this reason, feeding wild dolphins is banned in the USA.

Given that dolphins are large, powerful, wild predators, our desire to get into the water with them is surprising. However, when Georges, a wild solitary Bottlenose Dolphin previously recorded along the Brittany coast of northwest France and the nearby Channel Islands, came to southern England in early summer 2002, he had considerable human company. People packed into the water around him and his visits to the popular bathing beaches in Weymouth and neighbouring towns became a great cause for concern and difficult to manage. Despite the risks to him and his admirers, urging people to stay out of the water did not work and it was something of a relief when he moved on after a few months.

Like other cetaceans in close contact with human society, Georges was at risk of deliberate acts of aggression, potentially including retaliation if he had hurt someone. In fact, a year later, when Georges returned to the French coast, there were threats to kill him on the grounds that he had injured some swimmers. This issue, along with the risks of unacceptable disturbance to the natural behaviour of wild dolphins and disease transmission, is why WDCS and other organizations have opposed the development of commercial swim-with programmes. Instead they encourage people to watch

dolphins, including solitary animals like Georges, from the shore, where the opportunities allow.

The spiritually uplifting power of nature remains something to celebrate. Conservationists realize that lying behind their efforts must be an appreciation for the natural world and that this may be most strongly borne out of personal experience. It is argued that fewer people would see the real animals if there were no cetacean captives and that this would undermine conservation efforts. However, there are good alternatives to captivity and swim-with programmes. Wonderful documentary footage now reveals cetaceans living naturally in their real habitats and can even follow them deep below the waves in a fashion that would not have been possible ten years ago. In addition, artful use of computer graphics and technologies can be used to recreate versions of the marine world that we can walk through without getting wet. These alternative ways to encounter, enjoy and celebrate these animals do not come with associated welfare and conservation concerns.

BELOW   Atlantic Spotted Dolphins, off Little Bahama Bank, are typically friendly and inquisitive. Certainly they are are not 'camera shy'!

Watching them, watching you

The volcanic cone of the island of Pico, wrapped in low, skinny clouds, hung like an unearthly classical Japanese painting off one bow, as our sailing vessel motored gently along. The waters below were crystal clear, shafts of sunlight spearing down into the blue-black darkness. Someone pointed and shouted. As on many days that followed during our whale-watching trip, we were about to be joined by a school of small, brightly marked dolphins. They came hurtling through the water towards the boat, leaping high, then swam in small groups on either side of the keel perfectly matching our speed and 'bow-riding'. These were Common Dolphins, their characteristic white flanks shining in the sunlight. Some of the school stayed further away, at first paralleling the boat at a distance. Then, a few minutes later, a few other dolphins, perhaps a little more cautious, swam closer and amongst these were some small calves, swimming in the characteristic following position close to the side of their mothers' tails. One mother–calf pair, the younger animal about one-third the size of its parent but otherwise a perfectly formed replica, swam under the keel. The mother, still swimming swiftly, turned on her side and looked straight up through the water at us. A fraction of a moment later, the calf did the exact same thing. Clearly, we were watching them and they were watching us.

A few minutes later, the dolphins started to tire of us, peeled away in small groups and dropped astern. As its mother moved away, the small calf stopped paralleling its mother's movements for a few minutes and leapt remarkably high several times out of the water (albeit in a slightly clumsy and unco-ordinated fashion). The youngster gave the lasting impression that it was either showing off or trying to make some sort of comment to us (or perhaps both).

Extract from the author's journal during a whale- and dolphin-watching holiday in the Azores, 12th August, 2000.

The above was written during a trip where, in addition to several dolphin species, we saw pods of female and young Cachalots (a speciality of the islands), beaked whales (a rare phenomenon under any circumstances), turtles and sharks. The Azores have become one of several destinations around the world where, because of good access to spectacular cetacean species, whale-watching now provides a significant boost to the local economy. Worldwide some nine million people are now estimated to take part in commercial whale-watching trips in more than 90 countries. The scale of this new industry is also reflected by its value, with an annual global turnover estimated at US$1 billion.

BELOW   A Humpback Whale provides a spectacular breaching display in the Auau Channel, Hawaii.

Whale-watching opportunities range from short trips of an hour or so in small vessels to observe inshore cetaceans, typically dolphins, to entire holidays focused around whale-watching opportunities. The latter are typically boat-based expeditions lasting many days and passing through cetacean-rich waters. In between these two extremes are day-long trips in larger vessels that move swiftly out to deeper waters, probably in search of larger whales or offshore dolphins. You can go on organized whale-watching tours in the Pacific islands,

OPPOSITE   Underwater surfing: Atlantic Spotted Dolphins 'bow-riding' on the pressure wave in front of a sailing vessel in the Azores.
BELOW   A friendly face – an inquisitive Grey Whale calf in the San Ignacio Lagoon, Mexico.

Australasia, Japan, South Africa, Iceland, Greenland, Norway, Latin America, the Caribbean, North America (where it all began) and even in Antarctica and the Arctic, and many other places besides.

On balance, conservationists see well-managed whale-watching in a positive light as a significant way of introducing people to the animals, and many whale-watching operators are also increasingly involved in conservation. In some places they form the only regular monitors of the local cetacean populations. However, whale-watching needs to be conducted carefully, with the welfare and the best interests of the animals in mind. Unfortunately, there are certainly places where this is not the case and the intensity of whale-watching operations and the ways in which they are run raise welfare concerns and threaten local cetacean populations. For example, in some areas numerous whale-watching vessels target the local cetacean populations all day long. There are also growing concerns about the effects of this on the whales' natural behaviour and also the levels of noise that whale-watching vessels can produce in close proximity to the animals.

Various guidelines have been generated to try to control both whale-watching operations and the activities of leisure craft around cetaceans. These might usefully recommend a period of time, such as a third of daylight hours, when the animals should be left alone, or the provision of 'refuge areas' which boats cannot enter.

In some countries disturbing whales and dolphins is an offence and a combination of guidelines identifying best practice, but underpinned by appropriate national law, will probably be the best way to manage whale-watching in the future.

In the Mediterranean and Black Sea region, the scientific committee of ACCOBAMS (see page 143) has recently completed a set of whale-watching guidelines. These address the issues common to whale-watching activities everywhere, including:

- Identifying a distance that boats should maintain from cetaceans, unless the animals themselves decide to approach a boat more closely.
- Defining the length of the watching period for each boat and the number of vessels that should be close to the animals at any one time.
- Recommending that operators of whale-watching vessels are accredited by a licensing system and that skippers, crew and guides should be appropriately trained.

ABOVE   A Spinner Dolphin, with a small hitch-hiking remora fish attached, and its admirers on a whale-watching vessel in Hawaii.

Strandings and rescue

The live stranding of cetaceans on the seashore can be a spectacular and very moving event. On the positive side, strandings can be a rare opportunity to see such animals, although the presence of viewers needs to be controlled so that it does not stress the animals. On the negative side, what follows can be an exhausting and highly stressful battle that may take days and end in a heart-wrenching failure. Once stranded, these animals – which are so superbly mobile in water – are typically unable to do anything to help themselves and, without human intervention, the great majority perish.

There are a number of reasons why whales and dolphins strand. For example, certain shallow, gently sloping shores, with soft sandy or muddy bottoms, often featuring sand bars that extend out to sea, seem to be particularly prone to live strandings. This is the case in some parts of New Zealand and it seems that cetaceans find this kind of shore particularly confusing. It may be that their echolocation abilities work less well in areas of soft sediments and that stranding results primarily from navigational errors.

Generally speaking, inshore dolphin species, such as the Bottlenose Dolphins, seem less likely to strand alive than offshore species, which may be less able to navigate shallow waters successfully. Cetaceans that strand in the UK are often old and sick individuals, with illness impairing their naviga tional or swimming abilities. These

BELOW   A tragic mass stranding of Cachalots (Sperm Whales) in Oregon, US, in 1979.

malfunctioning animals frequently strand alone here. Elsewhere, live group strandings of what seem to be basically healthy animals are more usual. In New Zealand, group strandings are so common that both the government and voluntary organizations have developed rapid-response teams of trained rescuers who attempt, where appropriate, to return the animals to the sea – a process that has become known as 'refloating'.

Many of the species that commonly strand as groups, such as Pilot Whales and Cachalots, are also highly social. It may be that their dependency on the lead animals is so strong that they even follow their leaders when they make such mistakes or have become sick or wounded. There is also considerable evidence that when one member of a group is in trouble, the others will attempt to help it, even if this causes them to enter a dangerous situation. So, when one member of these tightly bonded groups strands, the other members often become stranded too.

Sometimes live mass strandings are the result of infectious disease. For example, in 1990, large numbers of Striped Dolphins came ashore in the Mediterranean

suffering from a measles-like morbillivirus infection. Disease outbreaks usually only affect a single species and can be diagnosed from their symptoms. By contrast, live mass strandings of mixed species groups often point to human activities. A series of highly unusual and fatal strandings in the 1980s in the Canary Islands seemed to result from naval exercises that advanced through the deep waters between the islands and the African coastline, driving beaked whales ahead of them. More recently, similar strandings of beaked whales, in the Bahamas, Greece and, again, in the Canary Islands, have been associated with the use of powerful military sonars.

In recent years, much consideration has been given to how to respond to stranded cetaceans and there are special rescue teams on standby in the USA, UK, Australia, New Zealand and elsewhere. These teams can respond rapidly to notification of a stranding and

ABOVE   A rescue in progress in Golden Bay, New Zealand, following the stranding of over 60 Long-finned Pilot Whales in 1993.

arrive armed with appropriate equipment. In fact, the most important response, when faced with one or more live cetaceans stranded on a shore, is always to get the appropriate experts into play as swiftly as possible. Untrained would-be rescuers may wound cetaceans by moving them inappropriately. Anyone attempting to rescue a stranded cetacean does so at his or her own risk. These are often large and very powerful animals that can carry disease and, even when stranded, they can thrash about.

Veterinary experts have developed techniques and criteria that can be used to evaluate the status of stranded animals. Time is a very significant factor. The longer a cetacean stays on the shore, the less likely it is to survive, and so the rapid summoning and arrival of expert help is vital. Posters advertising the contact details of rescue organizations are often prominently displayed at key sites on the coast and, in many parts of the world, the police and other rescue services will also know who to call for help. It is important to appreciate that in many cases the kindest course of action will be euthanasia. Many stranded animals are ill anyway and others may have become fatally wounded during the stranding, for example by encounters with rocks. In the case of some of the largest animals, it may be impossible to get adequately large lifting equipment close enough to help them.

In the USA most stranded cetaceans are moved into rescue centres, including candidates for euthanasia. Elsewhere, they are usually treated on the shore.

There is still a need for skill sharing between these two approaches. Worldwide there are many stories of marvellous rescues where, against all the odds, cetaceans have been returned to the sea. Often the animals seem to bond with their human carers. Pilot Whales, for example, sometimes seem to be comforted by contact with a particular person and show distress if that person leaves. This may mean that the person concerned will need to be specially catered for to allow them to stay with the whale for the duration – sometimes days – of the rescue.

Calves that are still dependent on their mothers present a special problem. Even if they are taken away and painstakingly cared for, they are unlikely to be able to survive in the wild afterwards; they will not have learnt all the essential survival and social skills from their mother that wild cetaceans need. However, it may be worth searching for the mother, as was recently shown in the UK by the reunion of a mother Harbour Porpoise, spotted in the surf, with her young calf, which was found stranded nearby a little earlier.

BELOW   Special 'rescue pontoons' have been developed by Project Jonah (a rescue organization) and form the sides of a stretcher used to lift and support stranded dolphins and whales, such as this Bryde's Whale in Whangerei, New Zealand. The rescue pontoons also support the animal in the water, helping it to regain its balance and use of its muscles, faculties that will have been impaired because of the abnormal pressures on the body during its stranding.

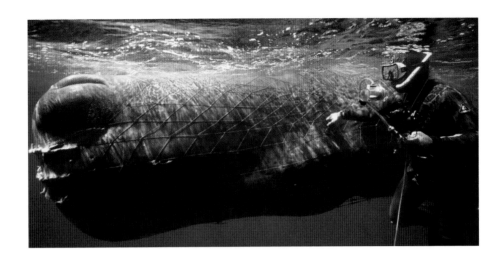

Chapter Four

THREATS IN A CHANGING WORLD

New threats to the world's whales, dolphins and porpoises are emerging as human impacts on marine ecosystems expand and diversify. In the past, the greatest danger to many whale and dolphin populations was from intentional hunting for meat, oil, baleen (see pages 120–122), ivory and other products. Now, in many localities, the major threats to cetaceans are incidental capture in fishing operations or habitat degradation.

Hunting

WHALING

The history of commercial whaling dates back to the 9th century when the Basque people (of north-east Spain and south-west France) started organized hunts for the North Atlantic Right Whales that used to come into the warm and relatively sheltered waters of the Bay of Biscay to rear their calves. When alerted by lookouts, these early whalers would row out and spear the whales, before towing the dead bodies to shore. Soon a wide range of whale products were being traded – meat, blubber, whale-bone (baleen) and oil – along the French and Spanish coasts and, by the 15th century, inshore North Atlantic Right Whales were becoming rare. The Basques responded by building bigger ships to hunt them offshore. In the succeeding centuries, other nations developed whaling fleets that became increasingly commercial, initially using the experienced Basques as their harpooners.

In the early 17th century, the Dutch established a whaling station in the Arctic at Spitzbergen, but this closed when the local whales became rare. Many populations of Right and Bowhead Whales in the Northern Hemisphere were decimated long before motorized vessels and explosive harpoons were introduced. The total disappearance of the North Atlantic Grey Whale population may also have resulted from these early whaling operations, even though the species is not mentioned in the whalers' records.

From the mid-1600s, Cachalots became the focus of a significant industry centred on Nantucket on the East Coast of the USA. After the American War of Independence knocked the North American whaling industry back, the British took the lead. As the whales became harder to find, competition increased and the French and American fleets started to attack British whaling vessels. Following the 1815 peace treaty with

PAGE 118   A Humpback Whale calf breaching in Icy Strait, Alaska.
BELOW   Humpback Whales and an old whaling vessel, which is now used for whale-watching in Hervey Bay, Australia.

Britain, the USA again expanded its whaling fleet and Nantucket had a renaissance. However, a sand bar limited the size of vessels that could enter the harbour and New Bedford, which could accommodate the larger whaling vessels, eventually became the main whaling port. Towards the end of the 19th century whale numbers in the Pacific whaling grounds were also falling, and the discovery of crude oil finally put an end to whaling using sailing vessels.

Modern commercial whaling began in 1864 with the invention of the exploding harpoon, by Svend Foyn, a Norwegian whaling captain. His design is very similar to that still in use today – with the harpoon fired from a cannon set in the bow of the vessel – and had a greater range than its predecessors. It penetrated the whale's body and exploded. The new steam-driven vessels that arrived during the 19th century also gave the whalers sufficient speed to catch the previously elusive fast-moving rorqual whales, such as Fin and Blue Whales.

The new larger, faster vessels could also now reach Antarctica, where large, hitherto untouched, whale populations could be exploited on their feeding grounds. Whalers developed a new technique that injected pressurized air into the whale bodies to stop them from sinking. In 1925 the first factory ship reached the Antarctic, enabling whalers to remain at sea for many months and to process and store whale products on board.

In the decades that followed, the slaughter in Antarctic waters was immense. The whalers focused on the biggest species, taking around 360,000 Blue Whales and almost 750,000 Fin Whales. The smaller species, including the Minke Whales – originally considered not worth hunting – became the primary targets after the far larger Cachalots, Humpback and Sei Whales were depleted. In total, more than two million whales were killed in the Southern Hemisphere during the 20th century.

ABOVE   A Humpback Whale – one of the great whale species previously prized by commercial whalers – passes by.

OTHER HUNTS

As noted in Chapter One, other cetaceans are still hunted in a number of places around the world, including Japan, where tens of thousands of small cetaceans are killed annually. The animals may be shot, netted, harpooned, or driven ashore and then killed. This last method is termed a 'drive fishery' and has also been the preferred method for centuries in the North Atlantic Faroe Islands, a self-governed territory of Denmark where the activity is called a grind.

In the Faroes, the whales are still driven into specially designated shallow bays by a line of noisy boats. Many of the whales are thus artificially stranded. The strong family bonds that exist in this species seem to ensure that others also stay around rather than heading out to sea and safety. The live animals are then secured by a hook or 'gaff', driven into their sides or inserted into their blowholes, and killed by being cut with a knife behind the blowhole. Although the hunt is described as traditional and meat is still distributed according to long-established rules, recent hunts have included the use of wet bikes and have been co-ordinated using mobile phones. Long-finned Pilot Whales remain the focus of the hunt and some 900 are still killed in this way each year. In addition, unknown numbers of dolphins are also taken, and stranded Northern Bottlenose Whales are also consumed.

BELOW An inquisitive group of Long-finned Pilot Whales 'spy-hopping' – peering around above the sea surface – in the Ligurian Sea Sanctuary.

Whither whaling?

The International Whaling Commission (IWC) was established in 1946 under the auspices of the International Convention for the Regulation of Whaling (ICRW). The IWC's objectives are defined as providing 'for the proper conservation of whale stocks and, thus, making possible the orderly development of the whaling industry'. Bearing in mind the date of the treaty, this dual objective – combining conservation with utilization – is not particularly surprising.

The IWC has jurisdiction over 'all waters where whaling is prosecuted' and all species of whales. However, the whaling nations typically contend that 'whales' should only mean the 'great' species taken in commercial hunts and that the IWC should not have authority over other, smaller, species. As a result, the IWC does not currently regulate the hunting of small whales, dolphins or porpoises.

For its first few decades the IWC did little more than oversee the decimation of one whale population after another, implementing a series of flawed management regimes. In 1982, in recognition of centuries of over-exploitation and lack of certainty about the status of whale stocks, the Commission imposed an indefinite moratorium on all commercial whaling. This came into effect in 1986, at which point several whaling nations gave up the practice in good faith.

However, certain 'loopholes' in the text of the ICRW have allowed some whaling for commercial purposes to continue. The convention allows any member to lodge a legal objection to decisions that would otherwise bind them. Norway did so and commenced 'whaling under objection' in 1992; today that country hunts over 600 Minke Whales annually. The ICRW also allows member nations to issue themselves with 'scientific permits' for the killing of whales for scientific research. Known as 'scientific whaling', this highly controversial provision is used by Japan to kill over 700 whales a year and, since 2003, has been used by Iceland. The ICRW was crafted at a time when there were no viable alternatives to lethal sampling, and scientific permits were originally issued for the study of a limited number of animals to inform the management of whale stocks. However, such data are no longer needed for whale 'management' and the IWC recently criticized the use of the 'scientific whaling' provision to facilitate commercial hunting.

BELOW   An Antarctic Minke Whale surfaces amid broken sea ice.

LEFT A spy-hopping Humpback Whale hangs vertically in the water.

OPPOSITE A yearling Grey Whale playing with seaweed off California. Elsewhere this species is in trouble and, in particular, there are grave concerns about the western Pacific population of Grey Whales, which is regarded as critically endangered and may only number 100 individuals. Their principal feeding ground is in Russian waters, off Sakhalin Island in the Okhotsk Sea. This area has recently been opened up for oil and gas development, prompting a statement of concern from many scientists worried about the implications of this development for the rare whales.

While the IWC prohibits whaling, another treaty, the Convention on International Trade in Endangered Species of Wild Fauna and Flora (CITES), prohibits international commercial trade in whale products. However, Norway, Japan and Iceland all hold objections, known as 'reservations', to the trade ban and, in recent years, Norway has exported surplus meat to Japan, Iceland and the Faroe Islands. Membership of the IWC is open to all nations and has grown significantly since its early days when the Commission was viewed as a 'whalers' club'. These days, the majority of its members have no interest in conducting whaling themselves. However, almost all members fall firmly into one of two camps – either supporting or opposing the resumption of commercial whaling, which is an ongoing, albeit informal, item on the IWC's agenda and which would require a majority of three-quarters to be approved. In 2003, at the time of the 55th annual IWC meeting, the IWC had 51 members, most of which oppose commercial whaling. However,

several pro-whaling nations have joined in recent years and the voting advantage may soon tip in their favour.

Although commercial whaling is prohibited, the IWC permits some groups of indigenous people – whose needs it formally recognizes – to conduct 'aboriginal subsistence whaling' for their local, non-commercial, use. These subsistence quotas are currently taken by indigenous people in the USA (Bowhead Whales), Greenland (Minke and Fin Whales), Russia (Grey and Bowhead Whales) and by the Bequians of St Vincent and the Grenadines in the Caribbean (Humpback Whales).

The work of the IWC is conducted through a number of working groups and committees, the most important of which is the Scientific Committee. This meets for two weeks prior to the annual Commission meeting and provides advice to the Commission (and the rest of the world) on the status of cetaceans. Its wide remit includes issues such as bycatch and the impact of environmental changes, including pollution.

Concerns about the health effects of high pollution burdens recorded in cetaceans have recently extended to the health of human consumers of cetacean meat and blubber. For example contaminants, such as mercury, have been found at high levels in cetacean products on sale in Japan.

WHALING AND WELFARE

The efficiency of a harpoon depends on where it hits the whale's body, but accurately hitting a partially submerged mobile target from the deck of a moving vessel is clearly difficult. Many whales are not killed outright, and may take several minutes – or sometimes hours – to die. In current commercial whaling and some aboriginal whaling operations, the harpoon is intended to detonate inside the whale, creating enough energy to cause sufficient trauma to the brain to kill it, or at least render it unconscious and insensible to pain. However, it appears that Japanese whalers may avoid the whale's head (the most appropriate target) because it would destroy samples needed for their 'research'.

The IWC has developed some simple criteria to judge when death has occurred, but recent studies indicate that these may not be adequate. For example, the fact that the animal is not moving and its lower jaw is slack (the primary IWC criteria) could hide the fact that the animal is paralyzed, but still alive and able to feel pain.

In the case of the small cetaceans, such as in the Faroe Islands, the whales and dolphins will be distressed from the moment they are trapped and, once stranded, are probably in pain. The process from the start of the drive to the eventual dispatch of the animals is inevitably long and it is not surprising that many feel that such hunts are unnecessary and unacceptably cruel.

In brief, strong arguments against whaling can be made from animal welfare and conservation perspectives:

- Many populations of whales are relatively small – either because they had been reduced by previous whaling activities or because, as is the case for many predators, they are naturally small.
- Satisfactorily determining and monitoring the population status of these animals over time is very difficult.
- The diving adaptations of the animals may make it difficult to determine if they are actually dead.
- Their sheer mass, complex blood systems and adaptations to marine life make it difficult to kill cetaceans swiftly and humanely.

Fisheries

BYCATCH

Cetaceans that become entangled in fishing nets usually die and, as fisheries have expanded worldwide, this 'bycatch' mortality has already brought several marine mammal populations to the verge of extinction. A report submitted to the IWC in May 2003 showed that from 1990 to 1999 the mean annual bycatch of marine mammals in US fisheries was 6,215 ± 1,415 animals – most dying in gill-net fisheries – roughly equally divided into cetaceans and pinnipeds (seals and sea lions). These statistics suggest that, globally, hundreds of thousands of marine mammals are being killed each year.

Some fishing methods are more problematic for cetaceans than others, particularly gill-net fisheries and offshore trawling. Gill nets are usually set on the seabed and left there until the fishermen return to retrieve them. The invention of strong nylon twines provided a cheap, tough and lightweight netting material and hugely increased the amount of gill netting being conducted. The strength of this new netting may also have increased the rate of cetacean entanglement. Cetaceans that tend to feed on the seabed, including the Harbour Porpoise and the Vaquita, are especially vulnerable to gill nets.

'Pingers', noise-producing devices attached to gill nets to make the animals more aware of them, are being advocated as a way to reduce entanglements. However, it is also possible that the high noise levels they produce might exclude the cetaceans from important habitat areas. These devices have reduced bycatch in some trials, but their use needs be very carefully managed and monitored before it can be concluded that they offer a solution to any particular bycatch problem.

Offshore trawling has also expanded. Huge nets are pulled behind a single large vessel or slung between two vessels to catch schools of fish swimming in mid or surface waters. In recent times, trawlers and the nets have grown significantly in size and some nets are as large

BELOW A Humpback Whale being released from a fishing net off the coast of Labrador, Canada.

as a cathedral. Offshore trawling has greatly expanded to the south of the UK since the 1980s and this correlates with hundreds of dead cetacean bodies coming ashore in midwinter on the French and adjacent English coastlines. The most frequently recorded species amongst the bodies is the Short-beaked Common Dolphin.

Before the expansion of the trawling industry in the Atlantic region, the most infamous incidental take of dolphins by a fishing activity was that of the eastern tropical Pacific tuna fishery. This fishery exploited the fact that tuna swim under the dolphins (for reasons that remain unknown), so when nets are set around the dolphins both they and the highly valuable tuna were captured. By herding the dolphins (perhaps by using smaller, fast-moving vessels) fishermen could even concentrate the tuna before the purse seine net was drawn tight. In the early days of this fishery many hundreds of thousands of dolphins died; worst affected were the offshore populations of the Pantropical Spotted Dolphin and the Eastern Spinner Dolphin. After a significant public outcry, the industry responded with various actions, including developing a 'back-down' technique that allows dolphins to escape.

Offshore drift nets are another particularly problematic fishing technique. These gill nets are supported by floats

ABOVE   Pantropical Spotted and Spinner Dolphins being released from a fishing net.
BELOW   A Pantropical Spotted Dolphin sadly trailing a monofilament fishing line.

OPPOSITE A Cachalot
with a hook and fishing line
in its mouth off the Bonin
Islands, Japan.
LEFT A Dall's Porpoise caught
in a driftnet in the Bering Sea.
BELOW A species once caught
in vast numbers in drift nets –
Northern Right-whale Dolphins
(California). Note the lack of
dorsal fin in this species.

at, or near, the water's surface, and drift with the currents. The invention of cheap (in this case, monofilament) nylon meant that drift nets of enormous size could be developed and, during the heyday of their use, nets extending more than 50 kilometres (over 30 miles) were deployed. The large-scale drift-net fisheries for squid in the North Pacific in the 1970s and 1980s caught a wide range of other species, including 2.4 million Blue Sharks, 19,000 Northern Right Whale Dolphins and many other species. In December 1991, the United Nations General Assembly passed a resolution calling for an end to large-scale drift netting on the high seas and a moratorium on their use came into effect one year later. More recently, the European Union banned drift netting in almost all its waters, and an extension of this has now been urged to protect the remnant population of Harbour Porpoises in the Baltic Sea. This is important because of the endangered status of this population.

PREY DEPLETION

Global demand for fish has doubled in less than 30 years and many fish stocks have been severely depleted. The effect of this on their natural predators is little understood. While there has been little correlation between prey reduction and cetacean population reduction, it would be very difficult to establish that a population decline was based on prey reduction when cetaceans are also being impacted by other factors,

including being killed directly in the nets of the fisheries.

Catches of most commercially targeted fish species, including North Atlantic herring and capelin, have declined significantly in the last two decades due to depletion of stocks. Declines in predator species, including some cetaceans and seabirds, have been linked with the decreasing North Sea herring stock. Although Harbour Porpoises feed on a wide range of fish species, the herring is especially important for them because it is particularly energy-rich, and trends in porpoise strandings seem to correlate with a decline in herring numbers.

CULLING AND CONFUSION

In recent years there has been much discussion about the potential impact of cetaceans on the fishing industry. Many pro-whalers claim that if whales eat fish, they must be consuming stocks that would otherwise be available to fishermen. Many others find this an overly simplistic and deeply flawed argument and suggest, instead, that cetaceans have become convenient scapegoats for devastating overfishing by commercial fisheries. Calculations have shown that, in a period of 15 years starting in the late 1980s, industrial fisheries typically reduced marine communities by 80 per cent and that the biomass of large predatory fish is only about 10 per cent of levels before industrial fishing began. The weight of evidence clearly indicates that humans, not whales, are to blame for declining fish stocks. Nevertheless, the argument that cetaceans and other marine mammals need to be controlled to 'save' fish stocks is gaining ground – despite the absence of scientific evidence.

Many whale species feed on non-commercial fish or invertebrates and others feed in polar feeding grounds distant from most fisheries' activities. Furthermore, the main predators of fish are actually other fish and not marine mammals. The argument that whales threaten global food security also overlooks the fact that today many whale populations are a fraction of the size they were before commercial hunting began, and that for millions of years prior to that abundant whale and fish stocks co-existed. Moreover, marine ecosystems are highly complex, and on the whole still poorly understood and there is little scientific consensus on the reliability of computer models to predict the impact of cetaceans on fish stocks or the implications of the removal of cetaceans.

There are some instances where individual marine mammals have become pests, for example, seals learning to raid fish farms and cetaceans stealing from fishing lines. These do not, however, amount to a justification for culling whole populations. Such localized problems may be addressed by non-lethal means, such as by using noises to scare the animals away or modifying the fishing technique being used. It is also worth noting that many fish consumers may not wish to purchase fish that have cost the lives of marine mammals.

Despite this, the principal objective of Japan's scientific whaling programme in the North Pacific in recent years is the study of the 'feeding ecology' of the whales. A resolution adopted by the IWC in 2000 urged Japan not to proceed with this lethal research and stated that 'gathering information on interactions between whales and prey species is not a critically important issue which justifies the killing of whales for research purposes'. Nevertheless, Japan continued with its research, ignoring an unprecedented request from 15 countries in 2000 to refrain from killing the whales. In August 2003, Iceland also began a scientific whaling programme to study the feeding ecology of Minke Whales in the North Atlantic. Ignoring criticism from the IWC's Scientific Committee, a rebuke from the Commission and the request of 23 countries not to proceed, Iceland killed 36 Minke Whales in the summer of 2003, and continued to take a few dozen individuals of this species in subsequent years.

Pollution

CHEMICAL POLLUTION

The development of synthetic chemicals that are highly resistant to degradation has resulted in the accumulation of these substances in the marine environment. Similarly, industrial by-products, including some naturally occurring substances, such as heavy metals, have also become significantly concentrated in marine systems.

Historically, the ability of aquatic environments to dilute polluting substances has been grossly overestimated. Many substances do not become evenly distributed throughout water bodies and thus diluted, but instead become associated with organic or other particulate matter. This contaminated material is then ingested by small organisms, such as plankton, which are ingested by crustaceans or small fish, which are eaten by larger fish, which are eaten, in turn, by marine mammals. In this way polluting substances are passed up the food chain and, at each stage, contaminant levels build up. The top predators, including many cetaceans, are the most heavily contaminated animals in the food chain.

Many synthetic chemicals are fat-soluble and accumulate in fatty tissues, and the fat-rich blubber of cetaceans renders them especially vulnerable to accumulating high levels of such pollutants. In addition, the metabolic pathways that help to break down toxic substances are different in cetaceans to those of terrestrial mammals and may, for some substances, be less effective. Of particular concern is the exposure of developing foetuses and calves, which occurs across the placenta and via mothers' milk. Their milk is exceptionally fat-rich and can be highly contaminated if the mother's blubber – which provides the fats from which it is produced – has accumulated high levels of pollutants.

Important fat-soluble pollutants include DDT (a pesticide) and the PCBs (a family of industrial compounds), which belong to the family of compounds known as the halogenated hydrocarbons (HHs). These have become infamous because of their abilities to disrupt normal body functions. They are particularly damaging to hormone or endocrine systems, which are important in maintaining the health of the individual, including its reproductive viability. In recent years, very high levels of HHs have been regularly reported in the blubber and other tissues of marine mammals.

BELOW   A school of healthy, fast-moving Striped Dolphins travelling at the surface of the sea off the Azores.

A few field investigations have started to confirm biological impacts on marine mammal populations, including the poor reproductive abilities of seal populations in the Wadden and Baltic Seas. Similarly, studies on Harbour Porpoises found stranded around the UK have shown a correlation between high contaminant levels and disease.

There has also been a recent spate of mass mortalities of marine mammals affecting Bottlenose Dolphins in the US, Harbour Seals in Europe, Striped Dolphins in the Mediterranean and several other populations. During these events, hundreds and, in most cases, thousands of animals have died and the proximate cause has usually been identified as a viral infection. However, there is considerable evidence to indicate that environmental problems made these populations unnaturally vulnerable. For example, seals that died in the 1988 European epidemic were, on average, more heavily contaminated than the survivors, and there is similar evidence for the 1990 Mediterranean dolphin mortality.

NOISE POLLUTION

Like chemical pollution, marine noise pollution can be difficult to detect and can disperse over wide areas, potentially with subtle but important consequences. At worst, very loud noises could physically harm animals in close proximity to the source. As described earlier, cetaceans are highly dependent on their sophisticated hearing, so any damage to their ears is of considerable concern. Powerful noises could also cause other tissue damage and lower levels might cause disruption of normal behaviour, possibly including the displacement of animals from important habitat areas.

Large vessels are usually loud and the global increase in vessel size and traffic has fundamentally changed the underwater noise profile of the world's oceans. Even the ability of the great whales to communicate with each other over huge distances may have been reduced by 'masking' noise. Another concern lies with the expansion of the oil and gas industry further out into the deep seas, including 'seismic exploration', which uses high-intensity sound to investigate the sub-sea rock strata in search of fossil fuel deposits. Military interests are also increasingly using loud noises, because as submarines have become stealthier, so detection systems using stronger sonars have been developed. In the US the deployment of Low Frequency Active Sonar (LFAS) by the military has become a prominent public issue and, following a court decision there in 2003, peacetime use of LFAS has been significantly limited. Similar powerful sonars are in development in Europe and their deployment is planned in the near future.

Over the last few years there have been strandings of beaked whales in Greece, the Bahamas and the Canary Islands associated with the use of powerful military sonars. Strandings of live beaked whales are relatively rare and these occurrences indicate that these deep-diving animals may be particularly vulnerable to loud noise.

BELOW The huge floating and decomposing body of a dead Blue Whale in San Clements, California.

Pathological studies of the bodies of the whales that were stranded in the Canary Islands have increased this concern by revealing evidence of a condition resembling the decompression sickness (the 'bends'), which is seen in divers who surface too quickly to cope with pressure changes. Scientists have suggested that exposure to loud noise could cause the whales similarly to ascend faster than their physiology is adapted to allow.

Seeking to address an issue that is only partially understood requires a precautionary response and WDCS recently conducted a major review of the noise threat. Its recommendations are summarized here:

- Attention should be given to the development of international laws to regulate marine noise pollution.
- An independent body should be established to initiate, promote, monitor and fund research into marine noise.

- All major developments in the marine environment, including those of an industrial or military nature, should be subject to full environmental assessment in terms of their input of noise pollution to the wider environment.
- All major developments should make a public commitment to mitigate their effects relating to noise, employ effective mitigation measures and develop alternative technologies to address this issue.
- Navies should seek to mitigate their noise-producing activities, avoid the deployment of powerful sonars and, ideally, develop a treaty that means that powerful sonars are not required.
- Marine Protected Areas (MPAs) and sanctuaries should be developed to take noise pollution, and its propagation beyond their boundaries, into account, including by the creation of protective buffer zones.

DEBRIS AND PATHOGENS

No seashore today is free of unsightly and sometimes dangerous plastic wastes and other 'macro-pollutants'. These materials, which are slow to degrade, are dumped from ships or discharged from land sources. Ingestion or entanglement in marine waste can kill marine wildlife, and dead cetaceans have been found with plastic items in their guts, including toys, cups, surgeons' gloves and bags. Whether they eat these items out of curiosity, as a play activity or just by mistake, the outcome can be lethal. For example, a dead Pygmy Sperm Whale was found a few years ago with a hard plastic ball blocking the exit from its stomach.

Bathing in sewage-contaminated water is now widely accepted as a human health risk and the same is probably true for the cetaceans. The potential for the exchange of diseases between humans and cetaceans is a controversial topic, but there are diseases that we have in common such as salmonella, candida and herpes infections. The discharge of raw sewage into the sea will

ABOVE Most marine pollution is invisible but, nonetheless, there is growing evidence that chemical pollution is having a negative impact on the health of marine mammals. In this picture, an inquisitive Hawaiian Spinner Dolphin plays innocently with a highly visible pollutant, a plastic bag – hopefully this led to no ill-effects. Unfortunately, however, cetaceans sometimes eat plastic objects and this can kill them.

OPPOSITE While some cetaceans may sometimes be attracted by vessels – as in the case of this bow-riding Dusky Dolphin in Kaikoura, New Zealand – vessel traffic and other human industrial activities can generate significant noise pollution.

increase the opportunities for coastal animals to pick up such infections.

Sewage discharges, along with fertilizer run-off from the land, also contribute to the 'hyper-fertilization' of the sea or 'eutrophication'. High levels of nutrients can cause unnatural blooms of plant plankton that exhaust the oxygen in the water and create 'dead' zones. Some of the plant plankton can also be toxic to humans and animals.

Habitat disturbance and destruction

Chemical and noise pollution, litter and sewage can all degrade the habitats of cetaceans, making them less suitable to support these animals. At worst, altering water flows, building docks or dams and creating marinas may change a habitat so profoundly that cetaceans, such as river and coastal dolphins, can no longer survive there. For species that range over wider areas, either on regular migrations or to exploit 'patches' of resources, the concept of habitat is more difficult to conceptualize, but such species still favour certain conditions along their routes and at their destinations. Consider the implications if the great whales, at the point when their energy stores are at their lowest, arrive at their polar feeding grounds to find that the annual spring bloom of plankton has not occurred. This may seem an unlikely scenario, but the extent of the ice cover at both poles appears to be changing; this, in turn, is affecting marine food chains.

Similarly, the breeding grounds of some highly migratory whales have been threatened by commercial development in recent years, including industrial developments planned for the Grey Whale breeding lagoons in Baja California, where conservation interests prevailed. However, more recently, the feeding area of the highly endangered west Pacific population of the Grey Whale in the Russian Okhotsk Sea has been opened up for oil and gas extraction, causing much concern for the whales' survival. In many coastal areas, aquaculture has become a significant economic activity. However, this growing industry can be a significant source of nutrient, chemical and even (where acoustic predator-scarers are deployed) noise pollution. The potential for fish farms to disrupt native wildlife, including precipitating conflict with predators that may come looking for an easy meal, is becoming increasingly apparent and needs to be borne in mind as the industry develops.

Even the popularity of cetaceans can result in them being excluded from favoured habitats. Over recent summers, a group of Bottlenose Dolphins has regularly entered certain popular bays on the English 'Riviera' coast of south Devon and Cornwall in the UK. This is a rich environment and probably provides them with a variety of prey species. However, their presence provokes attention from a flotilla of small leisure craft and impromptu commercial dolphin-watching operators. Frequently, vessels surround the dolphins throughout the daylight hours and some have even been seen to drive straight at them deliberately. It has been reported that the dolphins have been driven out of the bays, which may be decreasing the already diminished population's chances of survival.

Living on a changing planet

For 3,000 years the Ward Hunt Ice Shelf on the north coast of Ellesmere Island in Canada has been a permanent Arctic feature. In the summer of 2003 it fractured, a surface freshwater lake drained and large ice islands floated away, threatening shipping and oil platforms in the Beaufort Sea. Some scientists interpreted this as evidence of global warming; others were just as swift to dispute this. Whatever the truth, in this particular instance, there is good scientific consensus that climate change is in progress and the marine environment is altering at a faster rate than at any other time since the modern cetaceans appeared. There is real doubt that they can adapt quickly enough to keep up.

Pollution in the atmosphere – principally carbon dioxide – produces the 'greenhouse effect', where less of the incoming radiation from the sun is radiated back out into space. This creates rising temperatures in the lower atmosphere and at the planet's surface, disturbing a climatic balance that suits the forms of life currently inhabiting this planet. An increase in global temperatures of about 0.6 degrees Celsius over the past century has already started to affect weather patterns, sea levels and sea circulations. An increase of several more degrees is predicted over the next hundred years or so. The warming effect is not spread evenly over the planet and the polar regions seem particularly vulnerable. In Antarctica the temperature appears to have risen several degrees in recent decades and sea ice is reported to be far thinner than in the late 1950s.

Whale species that make long annual migrations from the poles to the tropics may be most at risk because they are dependent on finding certain resources at either end of their journey. They may have some ability to adapt if warm water bodies (suitable for rearing their calves) move or prey-rich areas decline or move, but we have no real idea to what extent this will be possible.

OPPOSITE   Grey Whales courting in Baja California. There is much surface activity and gentle manoevering by Grey Whales at mating time. BELOW   A Beluga Whale – one of the truly Arctic cetacean species – travels along a lead (an opening in the ice). Cetaceans that live year-round in polar waters, as well as those that regularly migrate here to feed, are at risk from the reduction in ice cover triggered by climate change.

Another problem relating to atmospheric pollution is the increased penetration to ground level of certain dangerous ultraviolet radiation, usually called UV-B. This is caused by 'holes' – in reality, thinning – in the gases in the upper atmosphere that would previously have reflected much of this harmful radiation away. The cancer-inducing problems relating to UV-B exposure are quite well known by the human population, but less widely appreciated is the fact that high exposure levels will kill smaller organisms, including plankton at the sea surface. As ozone thinning is concentrated at higher latitudes and its annual peak may coincide with the main plankton blooms in polar waters, it poses another threat to the primary food source of many whales.

RIGHT   Peale's Dolphin – only known from southern South America and the Falkland Islands, where this individual was photographed.
BELOW   The Arctic environment is rapidly changing: here a Grey Whale, amongst the ice in the Russian Arctic, raises its tail flukes.

Hot topics

To sum up, the main issues that will affect cetaceans in the future, particularly over the next couple of decades, are as follows:

- Whaling and whale culling The debate about whaling will undoubtedly continue to be a very hot topic, at least for a while. It is also important in a wider sense because how we ultimately decide to treat our marine mammals will have knock-on effects for other human–wildlife interactions.
- Marine management and bycatch The conservation and management of the marine environment, in particular fisheries, is clearly a huge challenge. This includes catches of non-target species, including cetaceans. In many parts of the world we have already stripped our terrestrial environment of its natural assets – only about half of the land surface can still be called 'wilderness' – removing the vegetation cover and many of the native animal species, typically starting with the big predators. We are already well on the way to doing the same in the seas. Reducing biodiversity and upsetting the balance of nature in this fashion cannot be in our best long-term interests.
- Noise pollution Loud noise sources in the marine environment are set to increase and the primary problem here is the lack of public comprehension. Since this is a highly technical issue in an environment where we see little of what is going on, this is hardly surprising. The question is how to change this situation in a timely manner.
- Identifying, understanding and protecting 'cetacean critical habitat' If we are to protect cetaceans, we must also protect those areas essential to their survival. Again, here we have a problem of public comprehension, based on the popular notions that marine cetaceans can travel widely to find their needs and will just move away and live successfully elsewhere if displaced. Defining their 'critical habitat' will be the first step; the next will be developing management regimes that really work.
- Managing whale- and dolphin-watching appropriately We need to take care that our desire to experience these animals does not become a threat to them.

Some simple common sense rules – not getting too close, giving them rest periods and refuges – should help.
Developing a better understanding of cetaceans is key to all conservation efforts.

ABOVE TOP An inquisitive Indo-Pacific Humpbacked Dolphin in shallow and busy waters in Australia.
ABOVE The distinctively marked backs of the rather chubby Commerson's Dolphins in Patagonia.

Chapter Five

CONSERVATION ACTIONS IN THE 21ST CENTURY

At the beginning of the 21st century, marine wildlife faces numerous threats, old and new – whaling, bycatch, chemical pollution, disturbance, climate change and so forth. Indeed, a single cetacean population may be faced by several threats simultaneously, some with complex interactions.

Addressing our impact in the alien environment of the sea – especially as we strive to extract more and more marine resources – is a major challenge. This short chapter outlines some of the measures being developed and implemented to ameliorate our negative influences.

Public opinion is vitally important and conservation organizations often ask their supporters to write a letter or sign a petition. Small actions such as these can persuade a policy-maker or politician to support conservation measures when a sufficient number of individuals act together.

Guiding principles

Conservation challenges change over time. No sooner is one threat to cetaceans addressed than a new one emerges. While commercial whaling has been the focus of concerns in the last century, in the current one, bycatch has taken centre stage as new, large-scale fishing technologies have proliferated. Similarly, the threat posed to cetaceans by loud noises in the marine environment has only really become apparent in recent years.

The guiding principles for cetacean conservation should be:

- Act in a timely manner.
- Base action on sound information wherever possible.
- When in doubt, act in the best interests of the animals/populations/species.
- Assume – if something unusual is happening to a cetacean population – that it may well have a human cause.
- Strive to assess all the interacting variables (natural and non-natural) when evaluating risk.

The International Union for the Conservation of Nature (IUCN) monitors the status of species worldwide and officially decides when one is endangered, thus helping to trigger appropriate conservation actions. More than 10,000 internationally recognized experts from over 180 countries volunteer their services to the IUCN's six global commissions, which include the Species Survival Commission and its Cetacean Specialist Group. In 2003, this group published an important new action plan for all cetaceans, encompassing a number of key conservation recommendations.

Conservation plans have traditionally focused on the species as the 'unit' that should be conserved. However, individual populations may also be genetically or behaviourally distinct, or facing particular local threats. For example, neither of the two species of Bottlenose Dolphin is thought to be imminently endangered (they are classified by the IUCN as 'Data Deficient') but several of their populations are small and in decline, such as the one in the Moray Firth in Scotland. The IUCN and other bodies are now starting to address the need to conserve populations as well as species.

Assessment of the status of the population is usually followed by an examination of the wider environment in which the population is found and this formal 'Environmental Assessment' (EA – also called Environmental Impact Assessment) has become a central tenet of marine conservation. In some cases, an EA can be a legal requirement before a potentially damaging activity, such as laying a submerged pipeline, proceeds. The EA process can also identify where information is lacking and prompt those who wish to proceed with industrial activities to consider carefully their own plans alongside other local or regional impacts and activities.

It is often very difficult to assign cause and effect in the marine environment categorically. In these situations the animals should be 'given the benefit of the doubt', and this cautious approach is enshrined in many environmental laws as the 'Precautionary Principle' or 'Precautionary Approach'. A simple concept that urges action even where there is uncertainty.

LEFT   Bottlenose Dolphins in South Africa leaping from a wave that they were previously 'surfing'. PAGE 138   Hector's Dolphins, which are endemic to New Zealand, are amongst the world's most endangered cetacean species.

Marine protected areas

The designation of 'nature reserves' on land is well established. These areas are managed to maximize their conservation potential, and human impacts are typically managed, minimized or eliminated. The same concept is now being applied in the marine environment through the development of Marine Protected Areas (MPAs). In the last two decades most of the MPAs that have been established around the world have been close to shore (within 12 nautical miles of the coast), making them easier to monitor and regulate. In recent years high-seas MPAs have also become an important conservation topic, but their management presents a new set of challenges.

To date, MPAs have typically been developed to protect biodiversity or breeding grounds for commercial fish species. In some areas they protect habitat used by particular species but, globally, very few MPAs specifically protect cetaceans. For this strategy to be effective, MPAs need to encompass critically important parts of the animals' habitat. For a species like the Vaquita – highly endangered and with a limited range – it is possible to define a protected area that would encompass its entire life cycle. This is far more difficult for the little-studied and wider-ranging oceanic species. In these cases efforts are made to identify critical habitat, which can include areas or 'ocean conditions' used for feeding, hunting, breeding, socializing, raising young, communication and migration.

When the deep seas to the north and west of Scotland

BELOW The breath of Orcas hangs in the cold air of British Colombia in Canada.

were opened up for fossil-fuel exploration (made viable by the development of floating production rigs), loud seismic exploration noises became an issue in the centre of prime whale habitats. Although the authorities were unsure of the implications, they decided to act in a precautionary manner. Guidelines were issued and the seismic vessels were required to have marine mammal observers on board. If a cetacean was sighted within a certain distance of the vessel, seismic surveying was not allowed to start. Whether this approach has been adequate can still be debated, but protective actions were at least initiated even though the threat was unclear.

One approach to MPAs is to create a representative network of areas across a region. The European Union is in the process of establishing such a network ('Natura 2000'). Only the Common Bottlenose Dolphin and the Harbour Porpoise are currently recognized in the relevant EU law as requiring the establishment of MPAs or, as they are called in Europe, marine Special Areas of Conservation (SACs). The reason why these two species, rather than any of the other 28 or so also found in European waters, were chosen is unclear, but may relate to their mainly inshore distribution. A recent UK court ruling, however, has shown that SACs should also be established in offshore waters.

The approach taken in the Ligurian Sea Cetacean Sanctuary, established in November 1999, is rather different. The Sanctuary comprises 99,000 square kilometres (38,224 square miles) in the western Mediterranean and is an area of high biodiversity where 13 different species of cetacean can be found. However, the primary reason for its designation is that it is an important feeding ground for Fin Whales. It is also the first high-seas MPA to be designated anywhere in the world specifically for the conservation of any cetaceans.

The IWC can also designate 'sanctuaries' but, so far, this has only meant that whaling is prohibited in these regions. This is an important distinction, as the designation of an MPA implies an attempt to mitigate all threats to the species being protected within the declared area. The Southern Ocean and the Indian Ocean are IWC Sanctuaries, but Japan made a formal objection to their designation and continues to whale in these areas.

A recent IUCN report showed that some 102,100 protected areas now exist, and that about 90 per cent of these were designated over the last 40 years. However, only about 4,120 are MPAs, covering less than 0.5 per cent of the seas and oceans, and only about 3 per cent of these currently consider cetaceans. Therefore, one of the important challenges in the coming decades will be to ensure that more MPAs take cetaceans into account in an appropriate way. For example, none of the UK's marine SACs cover Harbour Porpoises, despite the fact that they occur within the boundaries of many of the existing sites.

ABOVE   A Harbour Porpoise amongst the seaweed off the Norwegian coast.

The letter of the law

Whales and dolphins around the world are affected by a variety of legal instruments at three levels:

- National laws enacted by individual countries.
- Regional measures agreed by coalitions of countries with a mutual interest in a particular area.
- Global treaties.

Many countries now have national legal provisions that make killing or wounding cetaceans punishable offences. Increasingly these are being extended to also include disturbing cetaceans. The US Marine Mammal Protection Act (MMPA) is widely regarded as the most comprehensive national marine mammal legislation. It provides a legal framework for US citizens to abide by in all their interactions with marine mammals within US jurisdiction, ranging from boat behaviour around these animals to dolphin death rates in fisheries.

Two exciting cetacean-specific regional agreements have emerged in recent years:

- ASCOBANS (Agreement on the Conservation of Small Cetaceans of the Baltic and North Seas) in 1992.
- ACCOBAMS (Agreement on the Conservation of Cetaceans of the Black Sea, Mediterranean Sea and Contiguous Atlantic Area) in 2001.

Both agreements require member nations to act in the best conservation interests of cetaceans within their defined geographical scopes. ASCOBANS has recently helped to generate a series of recommendations to address fisheries problems as well as an action plan for the highly endangered porpoise population of the Baltic Sea. ACCOBAMS has similarly developed an action plan and, amongst many other things, is developing guidelines for the blossoming whale-watching operations in the region. More information on these important agreements can be found on their websites (see page 153).

ASCOBANS and ACCOBAMS were spawned from the Convention for Migratory Species (CMS), one of a few international treaty bodies to have become significantly involved in cetacean conservation in recent decades. CMS has a global membership and gained its 84th member country, Belarus, in September 2003. Its aims are threefold:

- To conserve migratory species and their habitats by providing strict protection for the endangered migratory species listed in Appendix 1 of the convention (including many cetaceans).
- To conclude multilateral agreements for the conservation and management of migratory species listed in Appendix 2 (such as ASCOBANS and ACCOBAMS).
- To undertake co-operative research activities.

CITES (the Convention on International Trade in Endangered species of Flora and Fauna) is another important global agreement and regulates cross-border trade in species between its 164 member countries. Trade is strictly controlled for species and populations listed on its appendices. CITES presently prohibits any trade in whale products, respecting the IWC's moratorium on commercial whaling.

BELOW   Fraser's Dolphins on the move. The first description of this species, which was based on a skeleton, was recorded in 1956. Its external appearance wasn't described until 1971.

Research and education

Suddenly, a Minke Whale breaks the surface over on the starboard side. It comes up for a breath, then another and another. We can see its sharply pointed nose and its sickle-shaped dorsal fin. Then with its tail arched, it dives away out of sight. On the still water, it leaves its 'footprint'; a series of spreading circles marking each place where it surged against the water's surface.

As we progress north, we spot one group of Risso's Dolphins after another. All are behaving spectacularly... we witness multiple breaching, head banging and tail slaps. It is as if each group of dolphins salutes us as we pass. One animal is particularly spectacular and, as he leaps into the air, he spins around on his long axis and he does this maybe a dozen times without pause. These are big grey animals with shining white bellies and tall dorsal fins...

Nor are the Risso's alone in this show, for White-beaked Dolphins also come out of nowhere to join in. Surging through the surface water they reach the front of the Neptun [our survey vessel] and start to bow-ride extravagantly... Then we turn out of the Minches, clear the Butt of Lewis and the low long ocean swell kicks in and the Neptun begins her characteristic wobble. Suddenly, the ship veers off to one side – we are meant to be heading in a straight line towards the survey block – and then all engine noise ceases.

Extract from the author's journal during a cetacean survey in the waters to the north and west of Scotland, 12th August, 1998.

Suffice it to say that the *MV Neptun* was soon able to restart her engines and we proceeded on our way. 'Whale counting' – which is what we were doing – has become a sophisticated science because some of the most important, and often contentious, questions relate to population size and whether it is increasing or decreasing. This is important in terms of conservation and also in the debates about whaling.

Whale counting is conducted from boats, and less often aircraft, which follow preset survey lines. Teams of expert observers record the cetaceans that they see and, where possible, identify them as a particular species and age group. They also record other information, including the whales' direction of travel and their distance and angle from the boat. These basic data are then fed into a formula which, using a number of assumptions – including how many submerged animals were missed on the track line – generates a population density and, where the data are adequate, an estimate. The accuracy of the raw data and the appropriateness of the assumptions

used in the calculations often raise questions about the validity of the population estimates.

Acoustic techniques are now being considered as a way of either augmenting the existing visual technique or, at some point in the future, replacing it. Other scientific methods have also been deployed to investigate population sizes. Very recently scientists have used genetic techniques to provide a new estimate of the number of whales present in the North Atlantic before whaling began. Their results indicate that populations were originally in the order of 240,000 Humpback Whales, 360,000 Fin Whales and 265,000 Minke Whales. These are far greater than the previous estimates calculated using whaling records, and some 6–20 times larger than the present-day population estimates. While these new estimates have been criticized, it is clear that placing too much emphasis on old whaling records to establish regional population sizes (and therefore potentially conservation goals) may be unwise.

The end of the Cold War has provided an unusual source of information about whales. Scientists have been given limited access to the US Navy Sound Surveillance System (SO-SUS), an ocean-spanning network of hydrophones originally deployed for listening for Soviet submarines. SO-SUS data from the North Atlantic have revealed much about the movements of Humpback, Fin and Blue Whales, including the discovery that not all of them make the regular north–south migrations that were anticipated. The hydrophones have also enabled the ultra-low frequency calls of the Blue Whales to be studied.

Internationally, we still need far more 'fundamental' research into the biology of cetaceans to provide a better understanding of their life histories, ranges and needs. We also need more 'applied' research, focused on particular issues, including evaluations of bycatch or the effectiveness of particular management plans.

Important themes in current cetacean research projects around the world include:

- Studies of population discrimination using genetics and morphology (anatomical features) to define the geographical limits and ranges of populations.
- Tracking, usually utilizing radio or satellite tagging to determine ranges and behaviours (the use of tags raises a difficult issue as to whether the invasive nature of the tag attachment is outweighed by the benefits of the information gained).
- Other behavioural studies where researchers monitor groups and sometimes known individuals over time.
- Independent observer programmes to monitor bycatch and target catch in fisheries.

ABOVE Who is studying who? – a dolphin researcher and an Atlantic Spotted Dolphin in the Bahamas.

OPPOSITE Whales and dolphins often leave their 'footprints' on the surface of the sea: distinctive circular areas of calm water where they have surfaced. Lines of these footprints can be left by whales or dolphins that have broken the surface several times to breath as they travel along. Frustratingly, whale researchers sometimes see the footprints but not the cetaceans – especially if they result from one of the shyer species.

BELOW A White-beaked Dolphin breaching off Iceland.

LEFT A transient Orca, which bears a radio tag, in British Columbia.
BELOW A school of Beluga Whales viewed from the air in Hudson Bay, Canadian Arctic.

Rapidly evolving camera techniques have greatly affected cetacean field research. Video recording allows behaviour to be captured in the field and later analyzed. Similarly, digital cameras offer the researcher the chance to take many photographs without using up lots of expensive film. 'Photo-identification' has become the primary tool of much behavioural cetacean research, including long-term studies of Orcas and Bottlenose Dolphins. This technique uses photography as a scientific means of confirming the identity of individuals which, in turn, allows relationships between individuals, their

distributions and other key factors to be studied. Nicks and notches on the dorsal fins, perhaps combined with marks on the back, are the usual means for identification. Typically, many photographs are taken in the field and identification is then corroborated by very close examination back in the laboratory. The accuracy of such identification can also be judged and scored.

Even after scientific data have been collected and analyzed, resulting conservation action plans still need to take into account the social, economic, political and cultural framework in which they must work. The key to success often comes down to having the necessary 'will' to change the human behaviour influencing a problem. In the case of cetaceans the problem may be well defined, but the solution can require multilateral international agreements for which there may be little political appetite.

For conservation actions to be successful they need to have public support. This means that our growing knowledge of cetaceans needs to be relayed in ways that are adequately engaging and informative. If people do not appreciate these animals, there will be no will to conserve them; no funds to support research; no one to develop and implement conservation plans; and no advocates to work on their behalf in the international forum – such as the IWC – where their fates will be debated. We need to ensure that young people know about, and appreciate, cetaceans. The future belongs to the young and they, in turn, will become the future of every other living thing on this planet.

There are many wonderful cetacean books aimed at younger people and numerous films and videos showing the life of these animals in the wild. The internet is of increasing importance and it is now possible to find websites dedicated to different species as well as those playing recorded whale songs. One of the most recent developments is a 'real time' website – one such system operates at Hansen Island, British Columbia – where underwater cameras and hydrophones wait for the local Orcas to pass by.

Conclusion

WHALES AND DOLPHINS IN A CHANGING WORLD

Humankind dominates this planet and exploits its terrestrial and marine resources aggressively, often with little thought about the sustainability of its actions. At the beginning of the 21st century, changes in the marine environment are accelerating. There are growing numbers of vessels and more marine-based industries than ever before; industrial fisheries are causing unprecedented fish-stock collapses; global warming-mediated changes are occurring on a grand scale; and more pollution, including loud noise, is permeating the seas and oceans.

The world of the whales and dolphins is changing. Humankind is the main cause of this, and we should take steps to conserve the cetaceans before they are overwhelmed by the consequences of our actions. They are beautiful, inspiring and often intriguing creatures. Our knowledge of them is still building, but already we know that many are highly intelligent and live in complex societies, and that they are an integral part of natural marine ecosystems. We also enjoy their company – even if this is sometimes to their detriment.

In order to conserve the cetaceans, we must address the following challenges:

- Persuading the pro-whaling lobby that whale populations are too vulnerable and valuable to subject them to the risk of commercial whaling.
- Developing fisheries management strategies that give due consideration to bycatch.
- Preserving sufficient good-quality habitat to enable cetacean species to survive.
- Curbing our enthusiasm for introducing new technologies until a thorough evaluation of their impacts on the environment has been made.

We also need to answer the following questions:

- Will our conservation efforts outpace the negative impacts we are having on them and their environment?
- Will humankind be prepared to look beyond economic considerations and bestow protective status recognizing cetacean 'rights'?

Let us hope that we can conserve the cetaceans as part of our natural heritage and find appropriate ways to coexist – for our sake as much as for theirs.

BELOW Boy meets dolphins – a meeting of minds?

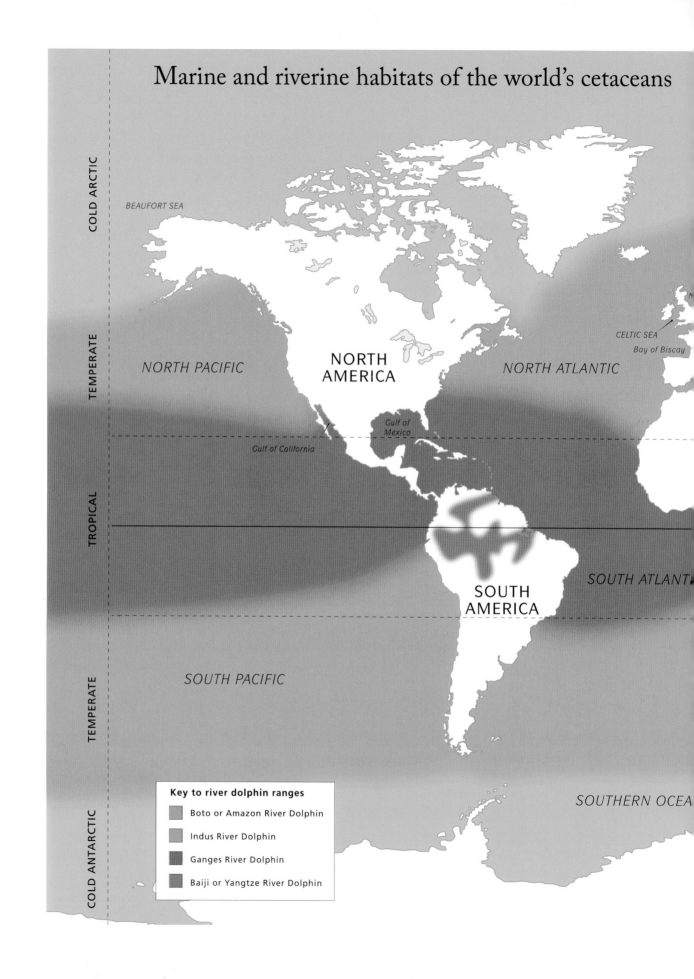

Marine and riverine habitats of the world's cetaceans

COLD ARCTIC

TEMPERATE

TROPICAL

TEMPERATE

COLD ANTARCTIC

BEAUFORT SEA

NORTH PACIFIC

NORTH AMERICA

NORTH ATLANTIC

CELTIC SEA

Bay of Biscay

Gulf of Mexico

Gulf of California

SOUTH AMERICA

SOUTH ATLANT

SOUTH PACIFIC

SOUTHERN OCEA

Key to river dolphin ranges

Boto or Amazon River Dolphin

Indus River Dolphin

Ganges River Dolphin

Baiji or Yangtze River Dolphin

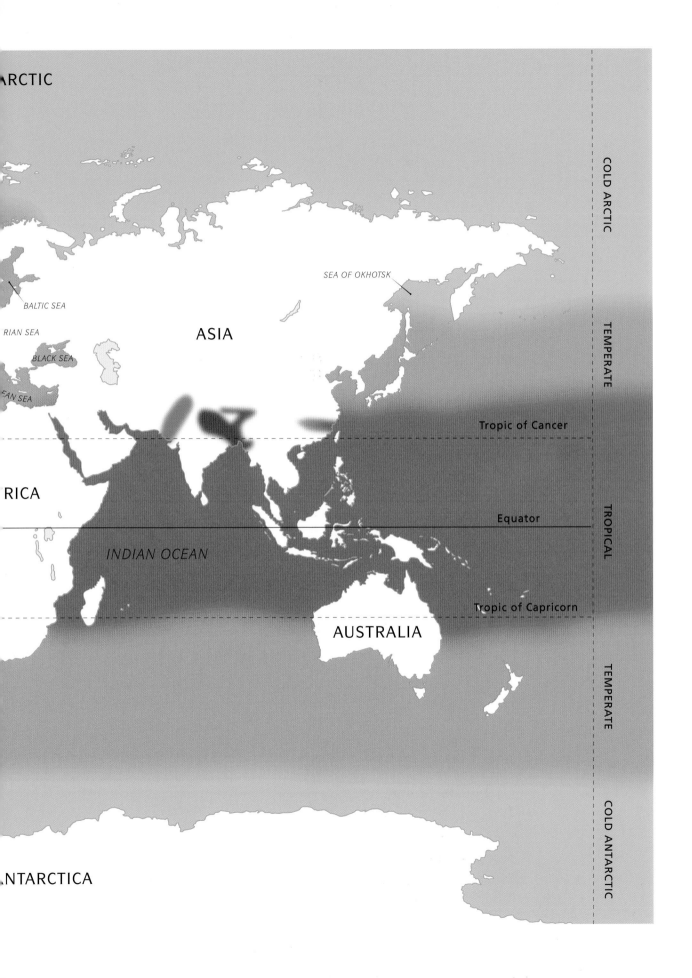

ARCTIC

BALTIC SEA

RIAN SEA

BLACK SEA

EAN SEA

ASIA

SEA OF OKHOTSK

RICA

INDIAN OCEAN

AUSTRALIA

NTARCTICA

Tropic of Cancer

Equator

Tropic of Capricorn

The cetacean species of the world

BALEEN WHALES (MYSTICETI)

The suborder Mysticeti contains four families: Eschrichtiidae (Grey Whale), Balaenidae (Bowhead and Right Whales), Neobalaenidae (Pygmy Right Whale) and Balaenopteridae (Roqual Whales). Between them, the families contain six genera and at least 14 species.

Bowhead and Right Whales

Bowhead Whale *(Balaena mysticetus)*
Southern Right Whale *(Eubalaena australis)*
North Atlantic Right Whale *(Eubalaena glacialis)*
North Pacific Right Whale *(Eubalaena japonica)*

Rorqual Whales

Common Minke Whale *(Balaenoptera acutorostrata)*
Antarctic Minke Whale *(Balaenoptera bonaerensis)*
Sei Whale *(Balaenoptera borealis)*
Bryde's Whale *(Balaenoptera edeni/brydei)*
Blue Whale *(Balaenoptera musculus)*
Omura's Whale *(Balaenoptera omurai)*
Fin Whale *(Balaenoptera physalus)*
Humpback Whale *(Megaptera novaeangliae)*

Grey Whale

Grey Whale *(Eschrichtius robustus)*

Pygmy Right Whale

Pygmy Right Whale *(Caperea marginata)*

TOOTHED WHALES (ODONTOCETI)

The suborder Odontoceti contains ten families, and includes dolphins and porpoises: Physeteridae (Cachalot), Kogiidae (Pygmy and Dwarf Cachalots), Monodontidae (White Whales), Ziphidae (Beaked Whales), Delphinidae (Marine Dolphins), Pontoporiidae, Platanistidae, Iniidae, Lipotidae (River Dolphins) and Phocoenidae (Porpoises). Between them, the families contain 40 genera and 71 species.

Cachalot

Cachalot (Sperm Whale) *(Physeter macrocephalus)*

Pygmy and Dwarf Cachalots

Pygmy Cachalot (Pygmy Sperm Whale) *(Kogia breviceps)*
Dwarf Cachalot (Dwarf Sperm Whale) *(Kogia sima)*

White Whales

Narwhal *(Monodon monocerus)*
Beluga *(Delphinapterus leucas)*

Beaked Whales

Arnoux's Beaked Whale *(Berardius arnuxii)*
Baird's Beaked Whale *(Berardius bairdii)*
Northern Bottlenose Whale *(Hyperoodon ampullatus)*
Southern Bottlenose Whale *(Hyperoodon planifrons)*
Longman's Beaked Whale *(Indopacetus pacificus)*
Sowerby's Beaked Whale *(Mesoplodon bidens)*
Andrew's Beaked Whale *(Mesoplodon bowdoini)*
Hubbs' Beaked Whale *(Mesoplodon carlhubbsi)*
Blainville's Beaked Whale *(Mesoplodon densirostris)*
Gervais' Beaked Whale *(Mesoplodon europaeus)*
Ginkgo-toothed Beaked Whale *(Mesoplodon ginkgodens)*
Gray's Beaked Whale *(Mesoplodon grayi)*
Hector's Beaked Whale *(Mesoplodon hectori)*
Strap-toothed Beaked Whale *(Mesoplodon layardii)*
True's Beaked Whale *(Mesoplodon mirus)*
Perrin's Beaked Whale *(Mesoplodon perrini)*
Pygmy Beaked Whale *(Mesoplodon peruvianus)*
Stejneger's Beaked Whale *(Mesoplodon stejnegeri)*
Spade-toothed Beaked Whale *(Mesoplodon traversii)*
Shepherd's Beaked Whale *(Tasmacetus shepherdi)*
Cuvier's Beaked Whale *(Ziphius cavirostris)*

Marine dolphins

Commerson's Dolphin (*Cephalorhynchus commersonii*)
Chilean Dolphin (*Cephalorhynchus eutropia*)
Heaviside's Dolphin (*Cephalorhynchus heavisidii*)
Hector's Dolphin (*Cephalorhynchus hectori*)
Long-beaked Common Dolphin (*Delphinus capensis*)
Short-beaked Common Dolphin (*Delphinus delphis*)
Pygmy Killer Whale (*Feresa attenuata*)
Short-finned Pilot Whale (*Globicephala macrorhynchus*)
Long-finned Pilot Whale (*Globicephala melas*)
Risso's Dolphin (*Grampus griseus*)
Fraser's Dolphin (*Lagenodelphis hosei*)
Atlantic White-sided Dolphin (*Lagenorhynchus acutus*)
White-beaked Dolphin (*Lagenorhynchus albirostris*)
Peale's Dolphin (*Lagenorhynchus australis*)
Hourglass Dolphin (*Lagenorhynchus cruciger*)
Pacific White-sided Dolphin (*Lagenorhynchus obliquidens*)
Dusky Dolphin (*Lagenorhynchus obscurus*)
Northern Right-whale Dolphin (*Lissodelphis borealis*)
Southern Right-whale Dolphin (*Lissodelphis peronii*)
Irrawaddy Dolphin (*Orcaella brevirostris*)
Australian Snubfin Dolphin (*Orcaella heinsohni*)
Orca (Killer Whale) (*Orcinus orca*)
Melon-headed Whale (*Peponocephala electra*)
False Killer Whale (*Pseudorca crassidens*)
Tucuxi (*Sotalia fluviatilis*)
Indo-Pacific Humpbacked Dolphin (*Sousa chinensis*)
Atlantic Humpbacked Dolphin (*Sousa teuszii*)
Pantropical Spotted Dolphin (*Stenella attenuata*)
Clymene Dolphin (*Stenella clymene*)
Striped Dolphin (*Stenella coeruleoalba*)
Atlantic Spotted Dolphin (*Stenella frontalis*)
Spinner Dolphin (*Stenella longirostris*)
Rough-toothed Dolphin (*Steno bredanensis*)
Indo-Pacific Bottlenose Dolphin (*Tursiops aduncus*)
Common Bottlenose Dolphin (*Tursiops truncatus*)

River Dolphins

Baiji (Chinese River Dolphin or Yangtze River Dolphin) (*Lipotes vexillifer*)
Boto (Amazon River Dolphin or Pink River Dolphin) (*Inia geoffrensis*)
Franciscana (La Plata River Dolphin (*Pontoporia blainvillei*)
South Asia River Dolphin (*Platanista gangetica*)

Porpoises

Finless Porpoise (*Neophocaena phocaenoides*)
Spectacled Porpoise (*Phocoena dioptrica*)
Harbour Porpoise (*Phocoena phocoena*)
Vaquita (Gulf of California Porpoise) (*Phocoena sinus*)
Burmeister's Porpoise (*Phocoena spinipinnis*)
Dall's Porpoise (*Phocoenoides dalli*)

Bibliography and resources

The author finds the following books particularly useful and usually keeps copies by his desk:

BOOKS

Byatt, A., Fothergill, A. and Holmes, M.
The Blue Planet – A Natural History of the Oceans
BBC Worldwide Ltd (2001).
A beautiful introduction to the marine world.

Carwardine, M.
Guide to Whale Watching in Britain and Europe
New Holland Publishers (2013).
An informative and practical guide to watching cetaceans.

Carwardine, M.
**Whales, Dolphins and Porpoises –
The Visual Guide to all the World's Cetaceans**
Dorling Kindersley (1995).
A fine and very informative guide.

Cohat, Y. and Colet, A.
Whales – Giants of the Seas and Oceans
Thames and Hudson (2001).
Helpful and pocket-sized.

Frohoff, T. and Peterson, B. (eds)
**Between Species: Celebrating
the Dolphin-Human Bond**
Sierra Club Books (2003).
An important collection of essays.

Hoelzel, A.R. (ed)
Marine Mammal Biology: An Evolutionary Approach
Blackwell Science Ltd (2002).
A scientific review of many important issues.

Mann, J., Connor, R.C., Tyack, P.L. and Whitehead, H. (eds)
**Cetacean Societies – Field Studies of Dolphins and
Whales**
University of Chicago Press (2000).
A landmark scientific volume describing modern field studies and cetacean behaviour.

Morton, A.
Listening to Whales
Ballantine Publishing Group (2002).
The powerful, fascinating and moving story of the author's life's work with wild Orcas.

Perrin, W.F., Wursig, B. and Thewissen, J.G.M. (eds)
Encyclopedia of Marine Mammals
Academic Press (2002)
Multi-authored and highly authoritative.

Reeves, R.R., Smith, B.D., Crespo, E.A. and
  Notarbartolo di Sciara, G.
**Dolphins, Whales and Porpoises 2002–2010
Conservation
  Action Plan for the World's Cetaceans**
published by the IUCN (2002).

Simmonds, M.P. and Hutchinson, J.D. (eds)
**The Conservation of Whales and Dolphins –
  Science and Practice**
John Wiley and Sons Ltd. (1996)

Whitehead, H.
Sperm Whales – Social Evolution in the Ocean
University of Chicago Press (2003)
Hal Whitehead's monograph on Cachalots – scientific but highly readable and fascinating.

Worldlife Library series
Colin Baxter Photography Ltd.
This is a series of well-written and well-informed books on cetaceans which includes Whales by Phil Clapham (1997).

Ridgway, S.H. and Harrison, R. (eds)
The Handbooks of Marine Mammals
Academic Press Ltd.
Of note are the volumes on the The Sirenians and Baleen Whales and The Second Book of Dolphins and Porpoises.

Also important are any of the excellent popular books by either Erich Hoyt or Mark Carwardine who, between them, have done so much to translate scientific knowledge into accessible and inspiring prose.

JOURNALS
Several scientific journals focus on marine mammal biology and issues, including:

Marine Mammals
Published by the Society for Marine Mammalogy.
A paper in the July 2003 volume is the source for comments about the status of Longman's Beaked Whale. The IWC (International Whaling Commission) also publishes many important documents, reports and books.

WEBSITES
Many websites now exist which provide detailed information about cetaceans, marine conservation issues, conservation organizations and agreements.

The Whale and Dolphin Conservation Society:
www.wdcs.org

The Environmental Investigation Agency:
www.eia-international.org/index_shocked.shtml

The Humane Society of the United States:
www.humanesociety.org/

Greenpeace:
www.greenpeace.org

**The IUCN (World Conservation Union)
Red List of Threatened Species:**
www.iucnredlist.org

The Convention on Migratory Species:
www.wcmc.org.uk

ASCOBANS
www.ascobans.org

ACCOBAMS
www.accobams.org

The International Whaling Commission:
www.iwcoffice.org

British Divers Marine Life Rescue:
www.bdmlr.org.uk/pages/main.html

Facts and figures

LARGEST
The Blue Whale is the largest animal ever to have inhabited the Earth. The longest individual to be recorded was a female measuring 33.58 metres (over 110 feet). Another female holds the weight record, weighing in at over 190 tons. Blue Whales are massive from birth; calves can be in excess of 5.9 metres (19 feet). The Fin Whale is the second largest animal ever to have existed.

SMALLEST
The Vaquita, a porpoise, is probably the smallest cetacean. The largest females reach only 1.5 metres (5 feet) in length and weigh little more than 45 kilograms (99 pounds). Their calves are tiny, just 0.7 metres (just over 2 feet) at birth.

OLDEST
Bowhead Whales are thought to be the longest-living mammal. Archaeologists dated a harpoon lodged in the blubber layer of a living Bowhead Whale in Alaska to the 19th century, and new ageing techniques estimate the age of one Bowhead Whale to be over 200 years.

FASTEST
Orcas and Dall's Porpoises compete for the title of fastest-swimming marine mammal. It is estimated each can reach maximum swimming speeds of over 55 kilometres/hour (34 miles/hour).

RAREST
The Baiji or Yangtze River Dolphin is the rarest cetacean in the world. If it is not extinct already there may only be a few individuals of this species left on the planet.

MOST TRAVELLED
The Grey Whale (East Pacific) and the Humpback Whales (Pacific and Atlantic) conduct the longest known migrations among the mammals. Each species covers in the region of 16,000–20,000 kilometres (10,000–12,500 miles) on its annual migration.

LOUDEST
Blue Whales produce one of the loudest calls of any animal at a low frequency (12Hz–0.2kHz) with a mean underwater source level of 188dB.

DEEPEST DIVING
The Sperm Whale holds the deep-diving record. They may reach a depth of 3,000 metres (9,850 feet) on their deepest dives, staying underwater for up to two hours.

LARGEST BRAIN
The Cachalot (or Sperm Whale) has the largest brain, weighing as much as 9.2 kilograms (20.3 pounds) – the size of a ripe water melon.

LARGEST REPRODUCTIVE ORGANS
The testes of the Right Whale are the largest of any animal. One individual's testes weighed 972 kilograms (2,143 pounds).

LARGEST 'BLOW'
When the Blue Whale exhales, its 'blow' (the water vapour in its exhalation) is visible on a calm day as a 10-metre (33-foot) high column.

FATTEST
The Bowhead Whale has the thickest layer of body fat of any animal. It can be 15–60 centimetres (6–23½ inches) thick.

LONGEST FINS
The pectoral fins of Humpback Whales can be up to 5 metres 16½ feet) long, the longest appendages of any mammal.

LONGEST BALEEN
The Bowhead Whale has the longest baleen plates. They are usually in the range 3.3–4.3 metres (11–14 feet) in length, but have been known to be as long as 5.18 metres (17 feet).

MOST POLLUTED
Toothed cetaceans are some of the most polluted animals in the world. An Orca found stranded on the Olympic Peninsula in North America had one of the highest levels of persistent pollutants ever found in an animal: 1,000 parts per million PCBs.

Glossary

Aquaculture: the cultivation of fish or shellfish under artificial conditions in the sea (or freshwater).

Breaching: leaping clear from the water.

Bycatch: the incidental capture of non-target animals in fishing nets – for example, dolphins in trawl nets.

Cephalopods: octopuses and squid.

Cetaceans: the collective term for whales, dolphins and porpoises – the members of the order of mammals Cetacea.

Cetologist: a scientist who studies cetaceans.

Copepods: small planktonic crustacean animals that are an important prey for some whale species.

Dorsal surface: the upperside of a cetacean, hence, the fin protruding from the middle of the back of most cetaceans is called the dorsal fin. The dorsal surface is usually darker than the ventral surface.

Habitat: the place where an animal, population or species lives and which contains all of its life requirements.

Hydrophone: microphone for recording sound underwater.

Lactating: feeding milk to young (lactation: the process of feeding milk).

Pair-trawling: a form of offshore trawling, where two boats tow one large net between them.

Pod: the term typically used to refer to a group of Orcas that regularly associate with each other and which are usually closely regulated.

Purse-seine net: an encircling fishing net that is drawn around the intended catch and then pulled closed at the bottom enclosing the fish.

Species-complex: some animals previously regarded as a single species (for example Orcas) are proving to comprise two or more species.

Spy-hopping: when a cetacean raises its head above the water surface to look around.

Ventral surface: the underside of a cetacean, including the belly and chest. The ventral surface is usually lighter than the dorsal surface.

Weaning: the process by which young progressively stop taking mother's milk.

Other books by New Holland

Index

Page numbers in *italics* indicate illustrations in the text.

Acknowledgements

The author is particularly grateful to Nicola Kemp and Vanessa Williams for their kind help in editing and otherwise improving this text. He also thanks the following for their comments or for helpful discussions: James Barnett, Philippa Brakes, Jo Clark, Sarah Dolman, Sue Fisher, Clare Perry, Margi Prideaux, Naomi Rose, Chris Butler Stroud, Dr Fernando Trujillo, Cathy Williamson and Simon Keith, who also helped to check many of the facts. Any mistakes are the author's and the opinions expressed here are his and not necessarily those of WDCS.

This text is dedicated to the Simmonds ladies – Nicky, Gillian, Vanessa, Sophie and Hannah – and to the author's nieces Stephanie and Anita Bryant.

UK £14.99